CONVEYORS FOR HANDLING BIOLOGICAL PRODUCTS IN-PLANT

Sabbas N. Asoegwu
Retired Professor and Former Head
Department of Agricultural & Bioresources Engineering
Federal University of Technology, Owerri.
Imo State, Nigeria

authorHOUSE®

AuthorHouse™
1663 Liberty Drive
Bloomington, IN 47403
www.authorhouse.com
Phone: 833-262-8899

Published by AuthorHouse 06/26/2023

ISBN: 978-1-7283-4597-0 (sc)

Library of Congress Control Number: 2020902437

CONTENTS

PREFACE

Conveyors are used in the production and distribution of everything we eat, drink, wear, live in, ride in, or in which we take pleasure - from the mine or the farm, through the factory or processing plant, through warehouses and distribution channels, and ending only with the ultimate user. This book, "Conveyors for Handling Biological Product In-Plant", will treat conveyors that move the food, feed, fuel, fur and fiber produced by agriculture and used as raw materials for the production of finished and semi-finished products in industries.

To ensure optimal design and selection of these conveyors, the properties of the conveyed biological products must be known. These properties are generally visco-elastic and the products are mostly non-free-flowing, thereby making the work of agricultural engineers in the design and selection of conveyors most difficult.

In-plant (industrial) transport or conveyance, which involves movement of materials within the manufacturing or production or processing plant, can be either continuous or intermittent. Engineers always settle for the continuous transport which is characterized by an uninterrupted product flow and by the use of a conveying device driven by mechanical or electrical force or by gravity.

In this book, an attempt has been made to summarize the methods of the design of conveyors: calculating the throughput capacity of, the forces acting on, and the power requirement of, each type of conveyor; taking into account all the relevant physical and mechanical properties of the product. These calculations as well as other design criteria are used to select the appropriate conveyor. Mathematical equations and problems are most often the best means of presenting and illuminating the principles of the design and selection of conveyors. Here, the author has used mathematical equations profusely. This means that an appreciable level of Mathematics and Physics has been assumed for whoever would be using this book. The various means available for horizontal, vertical or inclined conveyance covered in the book include belt conveyors, chain conveyors, bucket elevators, screw conveyors, shaking conveyors, pneumatic and hydraulic transport systems, gravity roller conveyors as well as fans and blowers. For the traction type ones, their maintenance checks were highlighted.

The approach used in writing the book makes it a good material for undergraduates and postgraduates in Polytechnics and Universities who are interested in the design of agricultural materials handling equipment. Sources of information including specific publications have been referenced in the text. Sufficient technical data have been provided in tabular form to enable the engineer in the design and selection of the conveyors. A completely new philosophy has been advanced to tackle the transportation question for biological materials in a processing plant.

Chapters 1 and 2 introduce conveying and conveyors, giving their several definitions and concepts, importance and a summary of their applicability in agriculture as well as the properties of the biological materials conveyed. The principles involved in materials handling were also highlighted. In Chapters 3 and 4, conveyors were classified, the general principles and steps to be taken when selecting them were discussed, and the general theory involved in calculating their throughput capacity and drive motor power was given.

Chapters 5, 6 and 7 discuss in individual details the traction-type conveyors in which the load moves together with the traction element (belt, chain, etc.). Their maintenance procedures are also included. The traction-less types (screw, shaking, pneumatic, gravity, roller, etc.) in which loads are transported by rotational, flowing or oscillatory movement are individually discussed in Chapters 8, 9, 10 and 11, respectively. At the end of the chapters on the specific conveyors, sample problems are given to enable the reader have a greater grasp of the theory discussed. In Chapter 12, fans and blowers used extensively in agricultural processing operations of drying, heating/cooling, ventilating/aspirating, cleaning, grading and conveying/elevating were discussed. These move air/gases, pure or mixed, with small sized agricultural particles. The differences between the axial and centrifugal types of fans were highlighted, while the fan laws were exhaustively treated.

S. N. Asoegwu
Department of Agricultural & Bioresources Engineering
Federal University of Technology
Owerri, 2020.

ACKNOWLEDGMENT

The author wishes to thank Professors **J. O. Duru and P. O. Okeke** of the Federal University of Technology, Owerri (FUTO) for their encouragement and reading the first draft of this book. The contributions of the Library Staff of FUTO are hereby acknowledged. The production of the book was made possible by the pain-staking typing-setting of **Miss Peace Elendu, Collete Opara and Blessing Monday Bassey,** at different stages of its preparation. The standardization of the format by **Engr. Uwen Ekpo, Miss Trudy Onyekwere** and **Mr Sasa Obrewovic** made the volume more readable. The contributions of my students over the past 30 years to my understanding of the methods of presenting the materials are immense. Finally, I gratefully acknowledge the patience and understanding of my wife **Dr. (Mrs) A. O. Asoegwu** and children, **Ugo, Chika and Ifeanyi**. The contributions of **Rev. Sr. MaryPaul Asoegwu** of the Daughters of Divine Love (DDL) Congregation are noteworthy and appreciated.

1

**IN-PLANT CONVEYING OF
BIOLOGICAL PRODUCTS**

1.1 DEFINITIONS:

A conveyor is a horizontal, inclined or vertical device for moving or transporting bulk materials, packages or objects in a path predetermined by the design of the device and having points of loading and unloading/discharge fixed or selective (Colijn, 1985).

A conveyor system is a common piece of mechanical handling equipment that moves materials from one location to another (McGuire, 2009). Conveyors may include all fixed and portable equipment for conveying biological materials/products between two fixed points with continuous or intermittent forward movement, but with a continuous drive. Conveyors are especially useful in applications involving the transportation of heavy or bulky materials. Conveyor systems allow quick and efficient transportation for a wide variety of materials, which make them very popular in the material handling and packaging industries.

Biological products describe all the food, feed and fiber, fuel and fur produced by agriculture including everything from fruits and vegetables to grain, hay, cotton, root, and tuber crops (Roth and Field, 1991).

The conveyor should change only the spatial location of the biological product, without changing its form or composition, except incidentally. These may occur in processes involving unit operations, transportation and storage (Pinches, 1958), utilization, marketing of biological products and even disposing of agricultural waste (Seferovich, 1958).

Materials handling are the arterial system over which the life blood of agricultural products flow on the farm and from the farm to the consumer. Another definition sees conveying as the picking up and putting down, moving of materials and products in any plane or combination of planes, by any means and it includes storage and all movements between processing operations except processing operations themselves (Kate, 2018).

1.2 CONCEPTS:

Research and development in materials handling may be guided and sustained by efforts to progressively eliminate those restrictions on the size of man's enterprise, or productive capacity, which are determined by how much he can move, carry or distribute in the working hours available to him. This makes the concepts of conveying two-fold:
- Labor saving, reducing or eliminating time consuming or onerous tasks; or
- The implementation of designs for the more efficient "flow of materials through the industrial/agricultural plant (in-plant)".

1

These concepts can be extended to include many fields agro-processing operations. Consider the volume and weight of agricultural products handled when processing them into industrial products: -

- Palm fruits to produce palm oil
- Palm kernel nuts to produce palm kernel oil
- Cocoa beans to produce beverages
- Maize grains to produce animal feed, corn flour
- Lump/cup rubber to produce crumb or sheet rubber
- Tomato fruits to produce puree, juice, ketchup
- Pineapple fruits to produce juice, jam
- Plantain fruits to produce fried slices, fruit bar
- Mango fruits to produce juice, jam, pulp, squash
- Orange fruits to produce juice, marmalade, squash, etc. (Asoegwu, 1989).

The performance of a processing plant (industrial or agricultural) is measurably affected by the efficiency of the movement of the materials from one unit operation to another, e.g., delivery of grains to a hammer mill and removal of the ground product in a grinding job (Henderson and Perry, 1955). There are a number of cases where an inefficient conveyor is used and where an efficient conveyor is poorly applied. These result from lack of knowledge of the characteristics of the conveyor and the material conveyed; from a poor integration of the conveyor and the material conveyed; as well as poor knowledge of the method for handling biological products. Agricultural engineers engaged in the design and selection of conveyors has an infinite variety of commercial conveyors available on the market today (Hoshimov et al., 2018). However, in some cases commercial conveyors must be altered and adapted and in some few cases new conveyors must be designed and developed. This book gives insight on how agricultural engineers can select or design conveyors appropriate to a given material handling situation based on the conveyor throughput capacity and power requirement.

1.3   IMPORTANCE OF CONVEYORS:

Conveyors have been defined as those machines that handle and move materials from one point to another. Every time a biological material is handled, something is added to the cost and nothing to its value (McGuire, 2009) since there is no change in its form or composition. This makes the design of agricultural materials handling equipment an important assignment for agricultural engineers. When properly designed, this equipment gives efficient, excellent and economical results. However, the problem of handling biological materials on farms goes far beyond tool invention and development. There are increasing concerns about flow processes which require chaining together several components into a system.

Large scale production and distribution in agriculture and industry will be impossible without adequate materials handling planning for improved efficiency and capacity of production. Internal transport of agricultural/biological products within a processing plant (in-plant) is intended to transfer and distribute loads within the plant from one process section to another. In modern plants, conveyors are the main type of in-plant or in-shop material handling facility, with trucks, carts and cranes used at times as intermittent conveyors.

In agricultural industries based on automatic flow-line production, conveyor transfer of articles from one process operation to another is most common. Process operations, such as cleaning, sorting, cooling, drying, mixing, packing, loading and unloading, etc. are performed on a moving conveyor. The conveyors mainly used in these in-plant processing operations of biological/agricultural products are discussed in this book. These conveyors are meant to:

- Establish and control production rate
- Ensure steady pace of production, and
- Are effective for increasing labour productivity and production output.

While material handling saves labour, reduces or eliminates time-consuming or onerous tasks, the implementation of its design ensures more efficient flow of materials through the plant. So, in modern plants, conveyors or materials handling machines may be employed as:

- Lightly productive transport machines for the transfer of goods from one process to another on lines of in-plant transport facilities and in some cases, in external transportation lines.

- Transportation units of powerful load transfer devices e.g. spreaders and charging/discharging machines

- Transfer machines in flow-line processes which establish, organize and control the pace of production.

- Machine and transfer devices in automatic processing flow-lines for the manufacture and treatment of parts and assembly components as well processing operations of agricultural products.

Conveyor systems are used widespread across a range of industries due to the numerous benefits they provide.

- Conveyors are able to safely transport materials from one level to another, which when done by human labour would be strenuous and expensive.

- They can be installed almost anywhere, and are much safer than using a forklift or other machine to move materials.

- They can move loads of all shapes, sizes and weights. Also, many have advanced safety features that help prevent accidents.

- There are a variety of options available for running conveying systems, including the hydraulic,http://en.wikipedia.org/wiki/Hydraulic mechanical and fully automated systems, which are equipped to fit individual needs.

1.4 APPLICATIONS OF MATERIALS HANDLING SYSTEMS AND EQUIPMENT IN AGRICULTURE

Conveyor systems are commonly used in many industries, including the automotive, agricultural, computer, electronic, food processing, aerospace, pharmaceutical, chemical, bottling and canning, print finishing and packaging, etc. (Thayer, 2017). Although a wide variety of materials can be conveyed, some of the most common include food items such as beans and nuts, bottles and cans, automotive components, scrap metal, pills and powders, wood and furniture and grain and animal feed.

In agricultural production and post-harvest processing, the materials handling systems and equipment are used in the following applications:

- Mechanical handling of solid fertilizers, liquid and gaseous fertilizers, farmyard manure and slurry.
- Chopping mechanism in forage harvesting which consists essentially of stationary knives, and an endless chain conveyor which forces the windrow across them. A chain-and-slat conveyor carries the crop to the rear, compresses it and later it is used to empty the wagon.
- Unloading of hay bales by conveyor-type unloading mechanism, and loading them by an elevator or elevator-conveyor system.
- Chopped hay is mechanically fed with augers or other conveyors to cattle and other animals.
- Augers, conveyor-belt, floating (chain and slat), and elevators are used in combine harvesters for different operations.
- Elevator chain, endless belt or roller chains are used in root harvesting machinery.
- Potato grader has roller inspection table and "chats" conveyor. Other graders may use roller or belt conveying surfaces.
- Roller feed conveyors are used in grading machinery for horticultural crops.
- In vegetable washing machines especially the continuous process ones, a final "picking-over" belt conveyor enables any blemished vegetables to be removed before packing.
- Hop-picking machines use conveyor chain and studded belt while pod-picker pea harvester employs a belt conveying system which collects pods, shelled peas and vine and delivers them to the threshing drum.
- The cabbage harvester has a rear-mounted endless belt conveyor-elevator.
- Seed cleaners, winnowers, separators, sorters and aspirating machines use pneumatic conveying systems. Some graders are used in corporate augers or belt conveyors.
- Feed mixers use auger conveyors, chain and slat inclined conveyors. Liquid feed mixing is carried out by hydraulic conveying.
- Endless-belt conveyor driers consisting of a series of well-fitting perforated plates or slats carried on chains and driven by sprockets are available. There are also rotary drum conveyor driers.
- Grain conveying can be done using pneumatic conveyors, bucket elevators and horizontal chain-and-flight type or even the endless-belt type. A general-purpose conveyor for moving grain at an angle is the high-speed auger. Horizontal sweep augers can also be used.
- In livestock production, augers, pneumatic, endless-chain and belt conveyors deliver meal, rolled barley, dairy nuts etc. to dispensing hoppers or mangers for pig-feeding, poultry feeding and filling of the feed dispensers in milking parlours.
- Pneumatic conveyors are used for handling grain. Forage blowers are for filling tower silos.
- Trench excavator employs an endless chain with bucket.

- Solid fuels are conveyed to burners using chain conveyor system.
- Chemical seed dressers make use of drag conveyors, screw conveyors and bucket elevators.
- Flax pullers are in the form of straight belt conveyors and rollers, curvilinear belt conveyors and rollers or belt and disc machines. The chain conveyor is used in sugar-beet harvesters.
- For a corn harvesting combine, the slat feeders may be either the endless chain or belt type conveyors.
- Silage in silos are unloaded pneumatically by forage blowers, supported by screw conveyor.
- Belt conveyors are used for handling commercial fruit sorting, grading, packaging and storage.
- Monorail conveyor lines are used in poultry processing.
- Screw conveyors, belt conveyors, hydraulic conveyors are used in the automated palm oil processing plants.
- Belt and roller conveyors are used in crumb rubber factories.
- Screw, belt, pneumatic, shaking conveyors are used in flour mills.
- Screw, pneumatic conveyors are employed in loading and offloading storage bins or silos.
- The electronic forage weighing machine between the silo unloader and the feeding equipment consists of a chain-and-flight conveyor at the feed end and supported on a load cell at the delivery end.
- Mobile silage feeders and mixer-feeders for feeding beef cattle has large-diameter horizontal augers as well as chain-and-slat conveyor.
- Mix-millers have auger meters and auto flow augers to convey ingredients to the hammer mill. Others have endless chain-and-slat-inclined conveyors with roller crushers. The mobile mill/mixer has a main feed-intake elevator; a pneumatic grain intake; a mixer feed auger; an intake auger to grain roller; a horizontal/inclined auger from a grain roller to mixer; a concentrate intake auger direct to mixer; a mixing and feeding auger.
- For controlled and measured rate of discharge from hoppers, feeders are installed at the hopper end. There are belt, apron, screw and vibratory feeders. Also, there are discharge aids which could be pneumatic-relying, vibrational-relying and mechanical methods (sweep-auger discharge or fixed auger discharge).
- Bucket conveyors/elevators are used to handle free flowing loads and find application in grain storage, at food-processing, and chemical plants, in building materials industry, and dockside installations for unloading ships.

There are numerous other situations in which conveyors can be put to use in agriculture.

1.5   MATERIALS-HANDLING PRINCIPLES:

Material handling is both an art and a science. The science nature is demonstrated by the fact that it has been possible for engineers to formulate a set of principles, developed and tested over time, which represent the rules that are to be followed in the design and selection of conveyors. The application of these rules or principles to particular problems and situations to enhance efficiency is the art. The selection of the conveying method depends upon the nature of application and on the type of material to be conveyed. However, to integrate the conveyors into a materials-handling system to enhance efficiency of operation, the following principles should be taken into consideration.

1.5.1 Materials Flow/System Principle:

This principle states that materials-handling efficiency is greatest when it approaches a steady flow, with a minimum of interruptions, backtracking and cross-hauling. The engineer's thinking should be that the movement of materials should approach a continuous rather than intermittent flow to eliminate batch processing and movement. Conveyor systems that tie several operations together are an excellent example of applications of this flow principle.

1.5.2 Mechanical Equipment/Automation Principle:

This principle states that efficiency and economy in materials handling results from using powered mechanical equipment in place of man power whenever practicable. Mechanical equipment offers the following advantages:

- Work done with equipment is generally more economical.
- A larger volume of work per operator is attainable, thus reducing the labour force.
- Greater speed in handling is obtained, and
- Safer working conditions result.

Where more mechanized mechanical equipment is to supplant existing mechanical equipment, the engineer should check and confirm if the existing/present equipment is used as efficiently as possible. The new equipment, if required, must be the simplest materials handling equipment applicable to the task. Special or custom-built equipment should be avoided because they are not cost effective.

1.5.3 Unit Load Principle:

This principle states that the efficiency in handling materials is increased as the size of the handling unit (load) is increased. A unit load is usually made up of several packages, containers, or items grouped together and can be handled and moved simultaneously or as a unit for unloading, transporting, stacking or loading operations or any combination of these operations. Unit loads reduce the number of individual handlings and trips required as well as the time required to complete the handling activity. Unit loads simplify inventory taking, increase storage capacity by permitting stacking and reduce the amount of spoilage resulting from picking up and dropping individual packages because of reduced number of handlings. Unit loads should be maintained for as long as possible. The longer they are kept together for all handlings and carried forward into other operations, the greater will be the savings and efficiency.

1.5.4 Balanced Handling Principle:

States that economy in handling materials is obtained when men and equipment are assigned in accordance with the minimum number of men required, and their work is so arranged that delay time and total man-hour requirements are kept to a minimum. A balanced crew and equipment is of utmost importance since it eliminates delays and bottlenecks and reduces the cost for the entire handling activity.

1.5.5 Flexibility Principle:

States that the greater the variety of materials-handling operations for which a particular type of equipment can be used, the greater is its value for materials-handling tasks. The ability for an equipment to be used and adapted to other tasks brings about greater utilization and reduced unit handling costs. The forklifts with different attachments for different types of loads and the farm tractor with power operated scoops, forks, etc. are examples. Portable conveyors have a lesser degree of flexibility.

1.5.6 Equipment Velocity Principle:

States that the economy of materials handling is usually increased as the velocity of the handling equipment is increased. The greatest value is obtained when the equipment is used to, but not in excess of, its full rated capacity. However, equipment should never be operated at such a speed or in such a manner as to create safety hazards. The optimum velocity at which the conveyor should be operated in order not to damage the material being conveyed or become hazardous to the operator and the environment should not be exceeded.

1.5.7 Equipment in Use Principle:

States that economies in handling materials are increased by decreasing the proportion of idle time to used time of the handling equipment. A unit of equipment is productive only when it is in use. Using poor loading or unloading methods that increase idle time also increase costs. Continuous loading at the rated feed rate is ideal for eliminating idle time and keeping the equipment in use most often.

1.5.8 Performance/Work Principle:

States that because several types of equipment are capable of performing the task, equipment selection should be based on economic performance rather than mechanical performance if maximum economy of time, manpower and money is to be obtained. Several materials handling equipment (farm tractors, conveyors, fork-lift, trucks, cranes, etc.) have their own characteristics with reference to speed, cost, maintenance of product quality, mobility, manpower requirements, etc. Therefore, the relative efficiency for performing a task should be measured in cost per unit handled. Cost implications of equipment performance should be of utmost importance when considering performance efficiency.

1.5.9 Equipment Replacement Principle:

States that economy in handling materials is increased by replacing handling equipment with more efficient equipment, when the savings affected are large enough to pay for the cost of replacement within a reasonable period of time. New developments in processing operations and new equipment and improvements in existing ones make it mandatory that the new equipment pays for itself within a reasonable period of time out of the savings it makes possible. The shorter the payback period, the more efficient the new equipment.

1.5.10 Safety Principle:

States that the productivity of materials handling equipment is increased with safe working conditions. The equipment selected for materials-handling operations should afford proper protection to the employee and to the equipment. Injuries are costly in terms of service costs, time lost by colleagues and management, liability on all employees. The safer it is to operate the equipment, the more savings on the total operations.

1.5.11 Facility Layout/Space Utilization Principle:

States that a good facility layout aids materials handling by smooth and direct-line flow of materials. Provision of adequate aisles or alley space for materials movement minimizes congestion and provides safe working condition. Space in material handling is three dimensional and therefore, is counted as cubic space.

There are other **Principles** to be taken into consideration: Environmental; Ergonomic; Life Cycle Cost; Planning; Standardization; for analysing and designing solutions to materials handling problems.

1.6 DESIGN PROBLEMS IN AGRICULTURAL MATERIALS HANDLING

Several problems make the design of agricultural materials handling equipment very challenging. Some of them are enumerated below:

- Agricultural materials are very diverse, of very irregular shapes and sizes. Most of them are alive and very perishable. These characteristics make their handling difficult with the tendency of designing an equipment for each crop. This makes the flexibility principle difficult to achieve.
- These different crops are produced using different types of agriculture in different regions of the world thereby complicating the way materials may be handled.
- In order to minimize cost, it is desirable to design materials handling equipment to handle more than one material, more so from the understanding that the equipment is used a small percentage of the time.
- The different types of agriculture are practised on both large and small farms thereby constituting problems of mechanizing of materials handling especially on small farms.
- No matter how much it is intended to manipulate some practices to aid mechanization, some old traditional practices will continue. While loose housing on a dairy farm is better from the stand point of materials handling, stanchion housing is still in demand and requires its own materials handling equipment.
- If a process can be made automatic (without an operator in attendance) and allows for continuous operation, a continuous or semi-continuous operation that takes small amount of load at a time should be considered and the equipment made lighter for less power consumption.
- However, if the process requires an operator in attendance, the equipment should be designed to handle large loads quickly to cut labour costs on the process, as the operator will be paid.
- As much as possible it is desirable to design all materials handling equipment to fit into a coordinated and integrated system which handles all materials for a given operation. Make the operation simple to use fewer man-hours.
- The facility layout and building for a particular coordinated system should fit the machinery outlay of the materials handling equipment for that system.
- Materials handling equipment that can handle more than one material should be designed for flexibility.
- Design a coordinated system that replaces existing field machinery that harvests and prepares the crop with one that puts out the material in a form adapted to the coordinated system.
- Design materials handling equipment to be self-feeding to save time, travel, labour and energy. "Don't handle it if you don't have to".
- Use engineering tools of work simplification, time and motion studies and workplace layout analysis for improving materials handling methods and equipment in a coordinated and integrated system.

CHAPTER 1 Review Questions

1.1   Define the conveyor; and in what processes are conveyors employed in agricultural processing operations?

1.2   What are the main functions of conveyors and what are they employed as in in-plant processing operations?

1.3   Name some of the applications of materials handling systems and equipment in agriculture.

1.4   Enumerate the materials handling principles and what they are meant to achieve.

1.5   Explain the mechanical equipment, materials flow, equipment velocity, performance, safety and facility layout principles of materials handling. What are their advantages?

1.6   Where are belt conveyors used in agricultural processing?

1.7   Where are screw conveyors used in agricultural processing?

1.8   Where are pneumatic or hydraulic conveyors used in agricultural processing?

1.9   Where are bucket elevators used in agricultural processing?

1.10   Where are roller conveyors used in agricultural processing?

1.11   Where are chain conveyors used in agricultural processing?

1.12   Where are oscillating or vibratory conveyors used in agricultural processing?

1.13   Name some problems that may be encountered in the design of agricultural materials handling.

1.14   What design consideration would you give if the equipment is to be used a small time for a continuous process?

1.15   How does farm size and diverse agriculture types complicate design problems?

1.16   In a coordinated process system, what design considerations would you give for equipment flexibility, man-hour reduction and layout adequacy?

1.17   What are the advantages of unit loads?

2

**PROPERTIES OF CONVEYED
BIOLOGICAL PRODUCTS**

2.1   INTRODUCTION:

In planning and designing a conveyor to handle biological products which are sensitive to treatments given to them, consideration is given to the nature of the products. Since every producer or distributor of agricultural products would want the commodities to reach the consumer in the good shape the consumer wants them, the nature of the product handled may be evaluated from:

- Its effect on the conveyor components
- Estimation and sizing of the power sources required to handle or convey the product.
- The effect of the conveyor on the product handled.

Many agricultural commodities are non-free-flowing materials. To transform them into free-flowing materials for the period when handling occurs requires the knowledge of the materials characteristics that encourage flow. Dry materials of almost symmetrical shape, such as sand and wheat, whose particles are held in place by friction on each other and without cohesion, are almost perfectly semi-fluids and are called free-flowing materials. However, with a cementing material between the grains, or a change in the characteristics of the material itself, cohesive forces are introduced and the material may become non-free flowing (Singley, 1958).

The distinction between free-flowing and non-free-flowing is thin and arbitrary. Free-flowing materials have no friction among its particles (a zero-degree angle of repose) and cannot be piled. However, grains such as wheat and shelled corn do repose by virtue of friction among the individual grains. Also, grain can quickly lose its flow ability when it contains a large amount of cracked grain, foreign material, and moisture (Barre, 1958).

The first principle of handling non-free-flowing materials is to exploit the **granular materials characteristics** that enable them to flow or to show a tendency to flow, as have been used with some success in the design of a silo that empties from the bottom and is used to store fibrous materials (Singley, 1957). With this in mind, agricultural materials to be handled can be characterized by their lumpiness, density (true and bulk), moisture content, angle of repose, particle mobility (fluidity), abrasiveness, strength, corrosion activity, stickiness, toxicity, explosiveness, self-ignition capacity and capacity for slumping and freezing. These and other properties/characteristics must be considered properly when designing or selecting the types and parameters of conveying or materials handling machines.

Materials are usually conveyed as unit or bulk loads. Most agricultural products are conveyed as bulk loads in their natural form (grains, milk, fertilizer, etc.). However, fruits and vegetables are better conveyed in containers as unit loads. The containers facilitate handling, protect the products, facilitate preparation of the product, provide a means of owner identification, and increase the flexibility of the operation (Levin, 1958). Although unit load handling has many advantages, many small expensive containers are still needed.

The introduction of lift equipment has revolutionized the handling of fruits and vegetables by replacing several small containers with one large one - a bulk box, or pellet box or bin. Such crops as onions, potatoes, pumpkins, pineapples, tomatoes, apples, pears, mangoes and oranges can be handled in bulk boxes without lowering the quality of the product.

The nature of a bulk load, described in terms of appropriate characteristics, is an essential consideration when designing or selecting equipment for its handling or storage. Description could be done on two levels.

- The behaviour of the material in its normal form as in compaction, flow and influence of moisture and electrostatic charging.
- The characterization of the constituent particles such as their size, density, hardness, shape and surface texture.

A "Materials Personality Test" (Table 2.1) helps to establish the behavioural features of a bulk solid. These descriptive terms are used for communicating information about the bulk solids. The designer of a handling system will require numerical parameters for use in characterizing the bulk solid, matching the system to the product, and assessing the likelihood of problems occurring or whether there has been a change in the product.

Table 2.1. "Materials Personality Test"

1	`Neurotic' materials have poor or too much flowability; they are sticky or tacky; they tend to pack or bridge.	They move awkwardly
2	`Sadistic' materials are abrasive, corrosive, toxic, explosive, hot.	They attack their surroundings.
3	`Masochistic` materials are friable, degradable, and can be contaminated.	They suffer from their surroundings
4	`Schizophrenic' materials are hygroscopic, susceptible to electrostatic charge.	They change their behaviour pattern

The properties of agricultural materials may be physical, thermal, chemical, mechanical (rheological), sonic, optic or electrical. In any given application one property or a combination of different properties may be important. In handling agricultural products, consideration is given mainly to both physical and mechanical properties of the product which will affect or influence their chemical and other properties.

2.2 PHYSICAL PROPERTIES:

The physical properties of agricultural products are some of the parameters which are important in many problems related to the design and function of a particular machine or to the analysis of the behaviour of the products under handling, processing and storage (Asoegwu, 1995; Reznicek, 1988; Mohsenin, 1970). These include size (volume), shape, surface area and other surface characteristics, weight (mass), density (unit, solid or bulk), porosity. In pneumatic conveying, consideration is given to the shape, volume and the surface area normal to the direction of flow. It should be noted that size, volume, area and density influence the aerodynamic characteristics of terminal velocity and Reynolds number. They are also important in the determination of the thermal diffusivity $\propto = \frac{k}{\rho}C$ in heat and separation of undesirable materials from the product.

2.2.1 Shape and Size and Volume:

Agricultural or biological products are generally irregular in shape and their dimensions are not uniform. This makes it difficult to define a single dimension which describes the sizes of particles. However, to represent the size of an irregular shaped particle by a single dimension, use is made of median size because the dimensions scatter around a mean value. Also, there are different size ranges for the same irregular product which can show different handling and conveying characteristics. Shape and size are important in the flowability of bulk products which influence conveyance. Irregular shaped particles in bulk load interlock and exert negative influence on the flow pattern. These three properties characterize a given product. For standard geometrical shapes, these properties are easy to determine. However, for the naturally irregular shapes of agricultural products, the determination of shape, size and volume is difficult. To determine the size of an irregular product, use is made of the following axial dimensions (Fig 2.1)

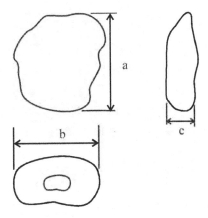

Fig. 2.1: Axial dimensions of irregular objects

13

where a = major diameter,   b = intermediate diameter, c = minor diameter measured mutually perpendicular to each other.

Based on the above, the following definitions are given:

i.   *Arithmetic mean diameter* $= F_1 = (a + b + c)/3$
ii. *Geometric mean diameter* $= F_2 = (abc)^{1/3}$
iii. *Square mean diameter* $= F_3 = [(ab + bc + ca)/3]^{1/2}$
iv. *Diameter of eqivalent sphere* $= D_e = (F_1 + F_2 + F_3)/3$

The shape of a product is defined using charted standards with numbers or descriptive terms (Mohsenin, et al., 1965) which have resemblance with geometrical bodies. Lemon has a prolate spheroid shape (an ellipse that rotates about its major axis). Grapefruit has an oblate spheroid shape (an ellipse that rotates about its minor axis). Other descriptive terms include ovate (egg-shaped and broad at stem end), obovate (opposite of ovate), oblique (lopsided - axis connecting stem and apex slanted), oblong (vertical diameter greater than the horizontal diameter), conic (tapered toward the apex like carrots and star-apple), round (approaching spheroid like *Irvingea gabonensis* or *Tetracarpidium conophorum*), cylindrical (approaching a cylinder like cucumber).

For prolate spheroid, volume, V and surface area, S are given by Eqn. 2.1

$$V = (4/3)(\pi ab^2); S = 2\pi b^2 + 2\pi(ab/e)sin^{-1}e \qquad \text{...2.1}$$

where a and b are respectively major and minor semi-axis of the ellipse of rotation and e is eccentricity given by Eqn. 2.2.

$$e = [1 - (b/a)^2]^{1/2} \qquad \text{...2.2}$$

For oblate spheroid, volume, V and surface area, S are given by Eqn. 2.3.

$$V = (4/3)(\pi a^2 b); S = 2\pi a^2 + \pi\frac{b^2}{e}ln[(1 + e)/(1 - e)] \qquad \text{...2.3}$$

For the frustum of a right cone, volume V and surface area S are given by Eqn. 2.4

$$V = h(\pi h/3)(r_1^2 + r_1 r_2 + r_2^2); S = \pi(r_1 + r_2)[h^2 + (r_1 + r_2)^2]^{1/2} \text{ ...2.4}$$

where r_1 and r_2 are respectively the radii of base and top and h is the altitude.

For some fruits like plantain and banana which have curved bodies, their shapes are not found in charted standards. However, Nwandikom and Asoegwu (1990) established some empirical equations relating some of the easily measurable physical characteristics to the others for use in the design of processing, handling and storage facilities for the plantain fruit. They found that the ratio of concave to convex length of plantain was 0.8, the fruit uniformity index ranged from 0.007 to 0.25. Some of the equations are given below:

$$L_2/L_1 = 0.44 + 9.83/L_1$$
$$FOC = \propto^o = 2Tan^{-1}(L_3/L_4)$$
$$FIC = \beta^0 = 2Tan^{-1}[L_3/2(L_4 + d)]$$
$$VF = 0.29d^2L_3 + 98.82$$
$$FI = 0.009d^2L_3 + 4.48$$
$$FI = 0.031VF + 22.06$$

where L_1 = finger convex length; L_2 = finger concave length; L_3 = straight finger length; L_4 = longest perpendicular distance from L_3 to concave; d = mid diameter of fruit; FOC = fruit outer curvature; FIC = fruit inner curvature; FI = fruit index; VF = volume of fruit.

2.2.2 Density

This property being a maturity index (Nwandikom and Asoegwu, 1990; Hightower, 1972) is a natural commodity constraint. The density of a substance is equal to the mass of the substance divided by the volume it occupies. It has to be considered during mechanical handling of agricultural products, in separation processes such as sedimentation and centrifugation, in pneumatic and hydraulic transport of powders and particulate as well as when mixing, storing and packaging particulates such as garri, flour, beans, rice, etc. Knowledge of the bulk density (mass/bulk volume) and the solid density (mass/solid volume) is essential for determining several important factors in the design of a system. However, the bulk density of a material will depend upon the solid density, the geometry, the size, the surface properties and the method of measurement. Because of the irregular shape of biological materials, determination of density is usually by the method of volume displacement. For a bulk material the average particle density is determined by dividing the mass of the material by the true volume occupied by the particles only (not including the voids). Density is also essential for selecting the type and throughput capacity of conveying machines, and determining the rated loads.

Bulk density ϱ_b is the mass of the material divided by its total volume (particles and voids) as given in Eqn. 2.5

$$\varrho_b = (M_{solids} + M_{voids})/(V_{solids} + V_{voids}) \qquad \qquad ...2.5$$

if ϱ_p = particle density and ϱ_f = density of fluid in the void spaces, then bulk density is given as Eqn. 2.6 where ε = porosity or void ratio

$$\varrho_b = \left(\varrho_p - \varrho_f\right)(1-\varepsilon) + \varrho_f \qquad \qquad ...2.6$$

For dry bulk solids the void spaces contain air of negligible fluid density ϱ_f compared to particle density, bulk density ϱ_b is given as Eqn. 2.7.

$$\varrho_b = \varrho_p(1-\varepsilon) \qquad \qquad ...\ 2.7$$

Knowledge of the bulk density ϱ_b and void ratio, ε of a product is essential in order to design storage vessels and conveying machines etc. However, in measuring this parameter, it is necessary to qualify any stated value with an indication of the condition of the material concerned, e.g. loose, or poured or packed or tapped.

However, volume and density of crops could serve as design factors for determining packing spaces, bio-yield and yield stresses, pneumatic and hydraulic handling conditions and for estimating quality (Nwandikom and Asoegwu, 1990; Nwandikom, 1990; Mohsenin, 1972). For conophor nuts, Asoegwu (1995) showed that its nut density was not only not significantly different among nut sizes, which is an important engineering design attribute, but that the nut has density = 0.877 kg/m^3 which is less than that of water.

2.2.3 Voidage (Porosity)

A bulk solid is a combination of particles and spaces between them. The percentage of the total volume not occupied by the particles is called the `voidage' or `void fraction' or `void ratio'. So, the porosity of a packed material is that fraction of the total volume which is occupied by air (volume of air/total volume) and it is affected by the geometry, the size and the surface properties of the material. When the container is tapped, the total volume decreases to equilibrium volume and gives rise to bulk density. The relationship between solid density, bulk density and porosity is given below:

$$voidage, \varepsilon = \frac{volume\ of\ voids}{total\ volume\ of\ particles\ \&\ voids}$$

$$\varepsilon = \frac{v_{voids}}{v_{solids} + v_{voids}}$$

$$\varepsilon = \frac{volume\ of\ air}{volume\ of\ bulk\ material}$$

$$\varepsilon = \frac{volume\ of\ bulk\ sample - true\ solid\ volume}{volume\ of\ bulk\ material}$$

$$\varepsilon = 1 - \frac{solid\ volume}{bulk\ volume}$$

Since solid mass = bulk mass, voidage is given as Eqn. 2.8

$$\varepsilon = 1 - (bulk\ density/solid\ density) = 1 - (\varrho_b/\varrho_s) \qquad \qquad ...2.8$$

The actual volume of solid particles being (1 - ε). Porosity is used for individual constituent particles. Voidage values range from 0.26 to 0.48 for agricultural materials.

2.2.4   Particle Size

Particle size gives some sort of average dimension across the particle. However, it is more appropriate to use qualitative terms (coarse, granular, fine) to describe the size of bulk materials whose particle sizes are within a certain range of values (Table 2.2). The particle diameter can describe mono-sized spherical particles. The average particle diameter can describe a mass of spherical particles of varying sizes. Image analysis has been used for the estimation of irregular particle size (Khanam et al., 2016; Schroeder et al., 2019).

Table 2.2   Qualitative terms and their size ranges

Descriptive term	Typical size range	Examples
Coarse (or broken) solid	5-100 mm	Coal, aggregates, etc.
Granular solid	0.3-5 mm	Granulated sugar (0.3-0.5 mm) Rice (2 – 3 mm)
Particulate solid Coarse powder Fine powder Super fine powder Ultra fine powder	 100-300 μm 10 -100 μm 1 – 10 μm < 1 μm	 Table salt (200-300 μm) Icing sugar (≡ 45 μm) Face powder Paint pigments

"An equivalent diameter" can describe an irregularly shaped particle. It corresponds to the diameter of a sphere that exhibits the same behavior as the particle under certain conditions. Volume diameter, d_v can be defined as the diameter of a sphere having the same volume as the particle and given in Eqn. 2.9.

$$d_v = \left(6V_p/\pi\right)^{1/3} = 1.241 V_p^{1/3}. \qquad \qquad \text{...2.9}$$

where V_p = volume of the particle,
Surface diameter, d_s is given as Eqn. 2.10.

$$d_s = \left(A_{sp}/\pi\right)^{1/2} = 0.564 A_{sp}^{1/2} \qquad \qquad \text{...2.10}$$

where A_{sp} = surface area of the particle.

i. Surface mean diameter d_{sm} is the diameter of a particle having a surface area A_{sm} equal to the average for all the particles in the mixture and given in Eqn. 2.11

$$A_{sm} = (1/N) \sum(\pi d_s^2); \quad d_{sm} = (A_{sm}/\pi)^{1/2} = \left((1/N) \sum d_s^2\right)^{1/2} \quad ...2.11$$

where d_s = diameter of sphere having the same surface area as the corresponding particle.

ii. Volume mean diameter, d_{vm}, is the diameter of a particle having the average volume, V_{pm}, for the mixture as given in Eqn. 2.12.

$$d_{vm} = \left(6V_{pm}/\pi\right)^{1/3} = \left((1/N) \sum d_v^3\right)^{1/3} \quad ...2.12$$

where d_v = diameter of a sphere having the same volume as the corresponding particle.

iii. Volume-surface mean diameter, d_{vsm}, is the diameter of a particle having a ratio of volume to surface equal to the average for the mixture (i.e, the diameter of a sphere having the same volume as the particle of average surface area for the mixture) is given in Eqn. 2.13.

$$d_{vsm} = \sum d_v^3 / \sum d_s^2 = d_{vm}^3/d_{sm}^2 = \sum(x/d_a)^{-1} \quad ...2.13$$

where x = mass fraction of particles passing through sieve aperture of size d_a.

2.2.5 Particle Shape

Shape and size are inseparable in a physical object, and both are generally necessary if the object is to be satisfactorily described (Mohsenin, 1970). The shape of the constituent particles in a bulk solid is an important characteristic as it has a significant influence on their packing and flow behaviour. Because agricultural products have irregular shapes, objective measuring indexes such as the roundness, the roundness ratio, the sphericity, etc., are used. For non-spherical particles, attempts have been made to establish the use of shape factors to indicate the extent to which particles differ from the spherical.

i. Sphericity is the reciprocal of the ratio of the surface area of a particle to that of a sphere of the same volume given in Eqn. 2.14.

$: \Phi_s = sphere\ surface\ area\ of\ same\ volume/particle\ surface\ area\ of\ same\ volume$

$$= \pi d_v^2/A_{sp} = (d_v/d_s)^2 \quad ...2.14$$

where d_v volume diameter and d_s = surface diameter. Also, sphericity can be expressed as Eqn. 2.15.

18

$$\Phi_s = \left(1.241 V_p^{1/3} / 0.564 A_{sp}^{1/2}\right)^2 = 4.841 \, V_p^{2/3} / A_{sp} \qquad \ldots 2.15$$

Other definitions of sphericity are as shown in Eqns. 2.16, 2.17 and 2.18.

$$\Phi_s = d_e / d_c \qquad \ldots 2.16$$

$$\Phi_s = [solid \ volume / circumscribed \ sphere \ volume]^{1/3} \qquad \ldots 2.17$$

$$\Phi_s = \left[\left(\pi/6\right)abc / \left(\pi/6\right)a^3\right]^{1/3} = \left(bc/a^2\right)^{1/3} = \left(abc\right)^{1/3}/a \qquad \ldots 2.18$$

where d_e = diameter of a sphere of same volume as the object and d_c = diameter of the smallest circumscribing sphere or the longest diameter of the object (Curray, 1951), and a = longest intercept, b = longest intercept normal to a; c = longest intercept normal to a and b.

ii.   Roundness is a measure of the sharpness of the corners of the particle (Eqn. 2.19).

$$Roundness = A_p / A_c = \sum r / NR = r / R \qquad \ldots 2.19$$

Curray, (1951) gave the least objectionable methods of estimating roundness as in Fig. 2.2 where A_p = largest produced area of object in natural rest position (Fig. 2.2b); A_c = area of smallest circumscribing circle; r = radius of curvature as defined in Fig. 2.2c; R = radius of maximum inscribed circle; N = total number of corners summed in numerator; r = radius of curvature of the sharpest corner in Fig. 2.2a; R = mean radius of the object. There are now simple and efficient algorithms to evaluate the roundness error from the measured points using four internationally defined methods: Least Squares Circle (LSC), Minimum Circumscribed Circle (MCC), Maximum Inscribed Circle (MIC) and Minimum Zone Circles (MZC) (Sui and Zhang, 2012).

Information on methods of sphericity measurement is quite poor. Some of the methods include: applying a classical 3-point method to measure spherical elements (Gleason and Schwenke, 1998); applying a confocal scanning optical microscope (Udupa et al., 1998); applying a special-purpose sphere interferometer in a combination with a stitching approach (Bartl et al., 2010); applying an optical microscope to measure spherical balls (Chen, 2007); and applying coordinate metrology in the area of measurements of spherical elements (Samuel and Shunmugam, 2003; Nafi et al., 2011). However, existing coordinate measuring machines do not offer a measurement accuracy (Janecki et al., 2012).

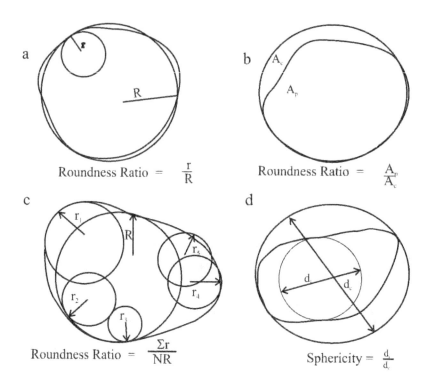

Roundness Ratio $= \dfrac{r}{R}$

Roundness Ratio $= \dfrac{A_p}{A_c}$

Roundness Ratio $= \dfrac{\Sigma r}{NR}$

Sphericity $= \dfrac{d_i}{d_c}$

Fig. 2.2: Illustrations for calculating roundness and sphericity

With average sphericity of about 0.91, Asoegwu (1995) suggested that compared to other fruits and nuts and seeds (Mohsenin, 1984), the conophor nut (*Tetracarpidium conophorum*) could be treated as a sphere for the design of its handling, processing and storage facilities. For *Irvingea gabonensis* nuts, sphericity was found to be 0.533 and its roundness was 0.735. It can be said to be more round that sphere (Asoegwu and Maduike, 1999).

iii.   Surface area is of considerable importance during handling and processing, especially of some parts of plant materials such as leaf area and surface area of fruits as well as of finely divided bulk materials such as catalysts, powders and paint pigments.
For the leaf of agricultural products, surface area has been related to the product of length and width of the leaf (Obiefuna and Ndubizu, 1979; Hughes and Proctor, 1981; Palit and Bhattacharyya, 1984; Asoegwu, 1988). The relationship is generally as follows:

$$S_L L = k \: x \: L \: x \: B \quad \text{or} \quad a + b \: L \: x \: B$$

where S_L= surface area of leaf; L= length of leaf; B= breadth of leaf; a, b and k = constants.

For fruits, surface area has been related to the fruit weight. Linear relationships have been established between them (Frechette and Zahradnik, 1965; Baten and Marshall, 1943) and given as follows:

$$S_f = a + bW$$

where S_f = surface area of fruit; W= weight of fruit; a and b = constants.

However, for eggs, Besch et al., (1968) established the following relationship:

$$S_c = kW^m$$

where k = a constant (4.56 - 5.07); m = 0.66; S_c= surface area of egg (in cm^2) and W = weight of egg in grams.

iv.   A specific surface is defined as the surface of pores in porous media exposed to fluid flow either per unit volume or unit weight. This parameter is required in studies of fluid flow and heat generation and transfer in storage.
For a single particle the volume specific surface is given as Eqn. 2.20.

$$S_p = (\pi d_s^2)/(\pi d_v^3/6) = 6/\varphi_s d_v \qquad \text{for } d_s = d_v \qquad \text{...2.20}$$

where d_s= surface diameter = d_v = volume diameter.

For bulk material, the specific surface in length squared divided by length cubed is given by the Carman - Korenzy equation as Eqn. 2.21.

$$K = P^3/5S_p^2(1 - P^2) \qquad \text{...2.21}$$

where P = porosity, K = permeability and S_p= specific surface. Darcy's equation gives permeability as Eqn. 2.22.

$$K = q\eta/A(\Delta p/L) \qquad \text{...2.22}$$

where K   =   permeability in m^2
       q   =   flow rate in m^3/s
       η =   viscosity of fluid in kg-sec/m^2
       ΔP   =   pressure difference across length of the pack (kg/m^2)
       L   =   length of porous pack in direction of flow (m)
       A   =   cross section area of flow (m^2).

The specific surface is inversely proportional to particle size. So, it can be deduced from a knowledge of the particle size distribution and used to indicate the "fineness" of a powder.

2.2.6 Particle Hardness

The hardness of the particles of a bulk product is valuable when designing a handling machine. It enables the designer to avoid undue erosive wear of the system components. The harder the particles, the more abrasive the products will be on the materials from which the handling machine is constructed. Hardness is measured using static indenters (Vickers, Rockwell or Brinell type) or using comparative measurements of hardness by simple scratch tests (Mohs hardness).

For fruits and vegetables which are harvested at a mature green unripe stage and left in storage for post-harvest ripening (e.g. plantains and bananas, mangoes and peas, and oranges, etc.) Their harvesting and handling must be at a stage of maturity to minimize damage. Asoegwu (1996) and Mohsenin (1984) identified firmness and hardness as important textural attributes in fruits and vegetables in connection with their loading, unloading and transportation to storage. They are the most commonly used engineering terms in quality evaluation of fruits and vegetables. From the engineering point of view, firmness is associated with tissue stiffness, modulus of elasticity, etc., as related to the design of containers for bulk transportation (Chuma et al., 1978) and fruit sorting machines (Delwiche, et al., 1989). Asoegwu (1996) also found that plantain fruit firmness decreased with increase in finger diameter, pulp/peel ratio and fruit water content. This shows the correlation between firmness and hardness. In ambient temperature storage, it was found that fruit firmness and fruit water content of plantains decreased with an increase in days in storage while pulp/peel ratio increased. However, at 6 days in storage, plantain fruit firmness was found to be 50 N and the pulp/peel ratio = 1.82; hard enough to withstand packing load and mechanical damage.

2.2.7 Cohesion and Adhesion

Cohesion is the resistance of a bulk solid to shear at zero compressive normal stress. The flowability of bulk solids is concerned with the forces of attraction or cohesion between its constituent particles; when these forces are low, flow is easy even under gravity. High inter-particle cohesive forces (caused by moisture or electrostatic charging) result in erratic material flow (e.g. wheat flour, cocoa powder and icing sugar). Adhesion describes the tendency of solid particles to stick to a containing surface of a channel or chute or hopper. It is useful in the design of systems involving the flow of bulk solids.

2.2.8 Angles of Repose and Surcharge

This is the limiting natural slope of the free surface of a bulk solid (Fig. 2.3a), i.e. the angle at which the loose bulk material begins to move. The poured angle of repose is the angle between the horizontal and the sloping side of a heap of the material poured gently from a funnel on to a flat surface (Fig. 2.3b). It gives a useful qualitative guide to the flow properties of a bulk solid, for example:

i. 25-30° Very free-flowing
ii. 30-38° Free-flowing
iii. 38-45° Fair flowing
iv. 45-55° Cohesive
v. >55° Very cohesive

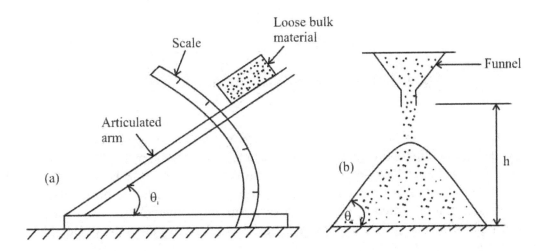

Fig 2.3: Determination of static angle of repose for bulk materials

The lower the angle of repose, the easier flowing is the product. The angle of repose indicates the contours of heaps of the material. It is required in order to determine the ullage space in hoppers or bins, the cross-sectional area of material transported on a belt conveyor, the surface topography of stockpiles.

The angle of repose depends on moisture content, temperature and lumpiness. It is influenced by the height of fall (h), by the filling capacity (t/h) and by the filling speed (v). It approximates the minimal angle of internal friction of the product $\mu_i = Tan\,\theta r$. In motion

$$\theta_s = (0.35 - 0.7)\theta_r$$

where $\theta_s = angle\ of\ surcharge$ and $\theta_r = angle\ of\ repose$

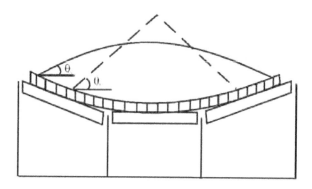

Fig. 2.4: Angles of repose and surcharge

The repose angle is important in materials handling and conveying since it affects the capacity of the belt conveyor and other bulk transfer devices and influences the height of material on a conveyor belt. When the belt conveyor moves, its movement causes the heap to spread a little forming the angle of surcharge. The angle of surcharge is the angle to the horizontal which the surface of the material now assumed while at rest on a moving conveyor belt. It is about 5° to 15° less than the angle of repose (Fig. 2.4). For most agricultural grain materials, the angle of repose $\theta_r \leq 28^o$ and the angle of surcharge $\theta_s \leq 10^o$ e.g. wheat (25°), barley (28°). Some of these materials range from very free flowing to free flowing (Table 2.3).

Table 2.3 Relationship between flowability, angle of surcharge and angle of repose

Flowability	Angle of Surcharge	Angle of Repose	Material Characteristics
Very free flowing	5°	0° – 20°	Uniform size, very small rounded particle, either very wet or very dry e.g. cement, sand, Amaranth and *Celocia* seeds, etc.
Free Flowing	10°	20°-30°	Rounded dry polished particles, of medium weight, e.g. whole grain beans, etc.
Average free flowing	20°-25°	30°-40°	Irregular, granular or lumpy materials of medium weight, e.g. clay, cotton seed meal, palm fruits, fruit vegetables, fertilizer, cup rubber, etc.
Sluggish	30°	> 40°	Irregular stringy, fibrous, interlocking, materials, e.g. wood clips, bagasse, sugar cane, cassava roots, etc.

2.2.9 Moisture Content

This is important in large and varied industries concerned with the handling, processing and storage of bulk solids including agricultural/biological materials. Moisture content can cause chemical change, deterioration of material quality and have dramatic influence on the flow behaviour of bulk solids by affecting the bulk density. Moisture content, W_m%, (Eqn. 2.23) is the ratio of the mass of water contained in the transported agricultural material, which can be removed by drying its sample at a temperature of $105°$ C to a constant mass of the dry sample on a dry basis (d.b.).

$$W_m = ((M_m - M_d)/M_d) \qquad \qquad ...2.23$$

where M_m and M_d are the mass of the moist and dry samples respectively.

2.2.10 Lumpiness

This is the distribution of particles between various sizes (particle size distribution). The uniformity of particle sizes of a bulk load is determined by the coefficient K_u (Eqn. 2.24) which is the ratio of the largest size of particles (lumps), a_{max}, , to the smallest size, a_{min}.

$$K_u = a_{max}/a_{min} \qquad \qquad ...2.24$$

For $K_u > 2.5$, bulk loads are ungraded ($a = a_{max}$); $K_u \le 2.5$, bulk loads are graded.

For graded loads, average lump size, a, is given as follows: $a = (a_{max} + a_{min})/2$

Lumpiness is considered when considering the dimensions of the load carrying elements of belts, beds, buckets etc.

2.2.11 Particle flowability, mobility or fluidity

This determines the cross-sectional area of the load on a moving belt or apron conveyor and the coefficient of lag of the load in the chute of a contoured - flight conveyor. The coefficients of internal friction f_i determines the angle of incline of walls and edges of conveyors and is associated with the angle of friction μ_f of the material by $f_i = Tan \, \mu_f$.

2.2.12 Abrasiveness

This is the capability of wearing or eroding the contacting surfaces of the elements of conveying machines. It depends on the hardness, shape and size of particles. All bulk loads can be characterized as A - non-abrasive, B - slightly abrasive, C - moderately abrasive; and D - strongly abrasive. Conveyors for transfer of abrasive loads should be designed properly to avoid quick wear of their elements.

2.2.13 Corrosion

This is the ability of some bulk materials to cause intensive corrosion (rusting) of steel elements. It calls for the application of special materials or protective coatings in conveyor design. This is important in hydraulic conveying of slurry in pipelines, washing machines with conveyor components. Corrosion behaviour of some construction materials used for agricultural handling and processing machines has been investigated (Asoegwu and Ogbonna,1999).

2.2.14 Explosiveness

Many bulk particulate solids like sugar, flour, cocoa, plastics, chemical, coal, wood, and etc., when dispersed in air to form a dust cloud, constitute a potentially explosive mixture. The risk of an explosion depends upon parameters such as product-to-air concentration and minimum ignition temperature and energy. The explosive mixture may be ignited by a naked flame, a hot surface or an electrical discharge. Explosions cause burns, deaths and serious structural damage. Three conditions must exist before an explosion can occur:

(i)   A suspension of combustible dust of explosive concentration,
(ii)   An ignition source (welding or cutting operations or smoking)
(iii)   Oxygen in sufficient quantity to support combustion.

Elimination of any of these conditions will certainly minimize the risk.

Flame is quenched as follows:

Upper limit $\rightarrow$ typically (2-10 kg product/m^3 air)
Lower limit $\rightarrow$ e.g. polyethylene (0.02 kg/m^3 air); coffee (0.085 kg/m^3 air)

For any concentration, the nature of a dust explosion is strongly influenced by the particle size of the material in the cloud. Particulate materials larger than 200 μm are not likely to cause an explosion. As the particle size is reduced, the given material becomes more hazardous and the consequences of an explosion more severe.

2.2.15 Self-heating (Self-igniting)

An ignition source can occur spontaneously as a result of self-heating, a phenomenon resulting from exothermic oxidation or decomposition of the product or bacteriological action. Because bulk solids have a very low thermal conductivity, the critical factor in self-heating is the rate at which heat is generated. Once the rate of self-heating exceeds the rate of heat dissipation a runaway situation occurs and an explosion is possible.

2.2.16 Slumping

This is the loss of mobility of particles on long storage (e.g. salts, lime, soda, cements, etc.). It grows with an increase in moisture content of the material and the pressure exerted on it. It is an annoying property especially in the design of hoppers etc., but it can be eliminated by employing various looseners (mechanical, pneumatic or vibrational).

2.2.17 Stickiness

This is the capacity of some bulk materials (clay, chalk) to stick to solids. This is avoided by selecting an appropriate shape of load-carrying elements (buckets) and supporting elements of the conveyor or by employing non-sticking coatings and various cleaning devices.

2.3   AERODYNAMIC PROPERTIES

The behaviour of agricultural products in a pneumatic conveyor is determined by their aerodynamic properties which include critical or terminal velocity, the coefficient of air assistance, and draft or drag coefficient. Size, volume, area and density influence these aerodynamic characteristics.

In a vertical stream of air, the agricultural products are acted upon by the gravitational force (its weight), G, and the resultant force, R, due to the air stream is given as Eqn. 2.25.

$$R = K\rho_a A(V_a - U)^2 \qquad \text{...2.25}$$

where K = coefficient of air resistance, ρ_a = density of air (kg/m^3), A = frontal area of product in direction of motion (m^2), V_a = velocity of air stream (m/sec) and U = velocity of the product (m/sec).

The resultant force, R, can be resolved into a horizontal drag force, R_d, and a lift or buoyant force, R_l, both of which are functions of area, density, viscosity and velocity. In the vertical direction, G and R act in opposite directions and the agricultural product may move down or up depending on whether G > R or R > G. It may also remain suspended when G = R, U = O and $V_a = V_{cr}$ = critical (terminal) velocity. Then, for G = R = condition of steady state is given by Eqn. 2.26 where G = weight of the agricultural product, N.s

$$V_{cr} = \sqrt{G/K\varrho_a A} \qquad \text{...2.26}$$

The critical velocity is determined experimentally using a drift classifier or in an aerodynamic tube wind tunnel. Once V_{cr} is got, the coefficient of resistance, K, and the drift coefficient, K_d, both of which depend on the shape of the product, its surface state and the medium in which it is located are got from the Eqns. 2.27, 2.28 and 2.28a.

$$K_d = 9.8K\varrho_a A/G = 9.8/V_{cr}^2 \qquad \text{...2.27}$$

The velocity of the air is calculated from the dynamic head of the air stream (h_d).

$$h_d = mV_a^2/2 = \varrho_a V_a^2/2 \ N/m^2 \qquad \text{...2.28}$$

where m = mass of 1 m^3 of air, kg, m $= \varrho_a$ = density of air (kg/m^3), V_a= velocity of air stream, m/sec given by Eq. 2.28a.

$$\therefore V_a = \sqrt{(2\,h_a/\varrho_a)} \qquad\qquad \ldots 2.28a$$

2.4   SOME BASIC CONCEPTS OF RHEOLOGY

Agricultural products are biological materials which are alive, constantly undergoing changes in shape, size, respiration and other aspects of life processes. Mechanical properties deal with the stress-strain behaviour of a material under applied force. Rheology considers the situation when the action of the forces results in deformation and flow which are time-dependent, e.g. creep, stress relaxation, viscosity.

2.4.1   Strain

Strain is the unit change in size or shape due to applied force as a ratio of the original size or shape given in Eqn. 2.29. It is dimensionless.

$$\varepsilon = \Delta L/L \qquad\qquad \ldots 2.29$$

Types of Strain

 (a) Linear Strain (tensile or compressive)
 (b) Axial, Transverse and Shear (angular) strain
 (c) True Strain $= n(\Delta L/L)$
 (d) Macro strain = strain over any finite gauge length of measurement which is large in comparison with interatomic distances.
 (e) Macro strain smaller than or comparable to interatomic distances.
 (f) Volumetric strain $= \Delta V/V = \varepsilon$

2.4.2   Stress

Stress is the intensity of a force on a body where this force component is acting in a given plane through that point. It is expressed in force per unit area as in Eqn. 2.30.

$$\sigma = Force/Area \qquad (N/mm^2) \qquad\qquad \ldots 2.30$$

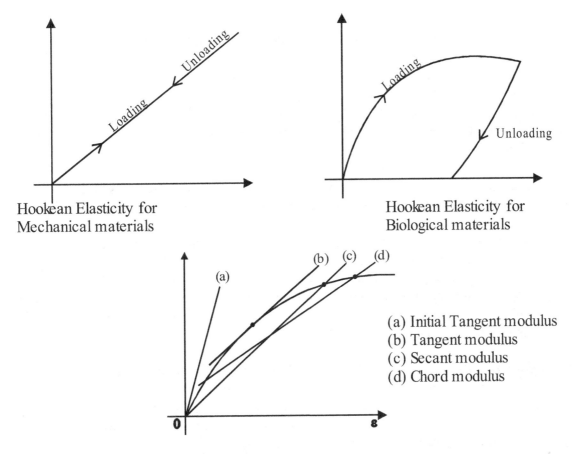

Fig. 2.5: Stress-Strain relationships in biological materials

(a)   Initial Tangent modulus - is measured by the slope of the curve at the origin
(b)   Tangent modulus - is measured by the slope or tangent at a particular point on the curve with specified stress and strain
(c)   Secant modulus - is measured by the slope of a line through the origin and cutting the curve at a specified point.
(d)   Chord modulus - is measured by the slope of the chord between two specified points on the curve.

Types of stress
(a)   Nominal stress (without regard to holes, grooves, fillets)
(b)   Normal stress (tensile or compressive)
(c)   Shear and Torsional stresses
(d)   True Stress (based on instantaneous cross-sectional area)
(e)   Principal stress
(f)   Fracture stress.

2.4.3 Elastic Limit

The elastic limit is the maximum stress a material can withstand or sustain without any permanent strain remaining after the release of the stress causing the strain. Mechanical materials experience Hookean elasticity where, within the elastic limit there is a direct relationship (correlation) between stress and strain. In biological materials, the stress strain relationship (Fig. 2.5) is curvilinear, producing residual strain when the load is removed. So, cognizance is not taken about modulus of elasticity here, but various terms are used to give the measure of rigidity (as given in Fig. 2.5 above).

2.4.4 Characteristics of Force-Deformation Curve for Agricultural Products.

1. **Deformation**, a vector quantity, is the relative displacement of points within a product as a result of applied forces (uniaxial compression or tension). It is accompanied by a change of volume (resulting from isotropic stress or hydrostatic pressure) or shape (resulting from non-isotropic stress, e.g. shear stress).

2. **Proportional (Linear) Limit** is the maximum stress a bio-material can withstand and still obey Hooke's Law. In the Fig. 2.6, O-L is the elastic region and L is the linear limit. This portion is usually very small for agricultural materials.

3. **A bio-yield point** is the point in the curve at which there occurs an increase in deformation without change or decrease in force. The presence of this is an indication of initial cell rupture of the bio-material. It may occur at any point beyond the linear limit like at point Y where the curve deviates from the initial straight-line portion. This rupture is usually internal (within the material) and not visible to the naked eye.

4. **Rupture point** is the point on the curve (like R) at which the material ruptures or fails under load causing puncture of shell or skin or cracking or fracture of the bio-material noticeable to the naked eye. While a bio-yield point denotes failure in the microstructures of the material, the rupture point is in the macrostructures of the specimen. How soon after the bio-yield point the rupture point occurs determines the toughness or brittleness of the bio-material.

5. **Stiffness or rigidity or apparent modulus** of a non-linear, stress-strain, behaviour, as in agricultural materials, are defined in terms of initial tangent modulus, secant modulus or tangent modulus. It is used synonymously with modulus of rigidity. Stiffness is directly proportional to the rate of loading (application of force).

6. **Toughness** is the amount of work or energy required to bring about rupture in a material. It is measured approximately as the area under the force-deformation curve when the material is loaded to rupture. It is given in terms of energy/unit volume (J/m^3) and specifying the loading surface used.

7. **Elasticity and Plasticity**: Elasticity is the capacity of a material for taking up elastic or recoverable deformation. Plasticity is its capacity for taking up plastic or permanent or irrecoverable deformation. For agricultural products all finite deformations tend to be made up of both plastic and elastic deformation.

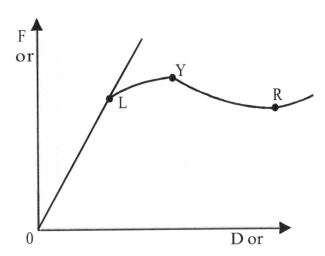

Fig. 2.6: Generalized force-deformation curve for agricultural products

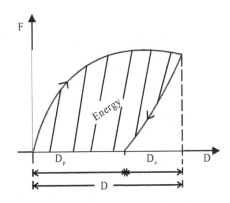

Fig. 2.7: Hysteresis in agricultural products

When bio-materials are loaded and unloaded within the proportional limit, hysteresis (Fig. 2.7) occurs and not all deformation is recovered. The ratio of the recovered deformation to the total deformation gives the degree of elasticity (Eq. 2.31).

$$D_e/D = D_e/(D_e + D_p) \qquad \dots 2.31$$

31

where D_e = elastic or recoverable deformation, D_p = plastic or residual deformation, D = total deformation.

8. **Mechanical Hysteresis** is the energy absorbed by a material in cycles of loading and unloading. It is evaluated as the area between the loading and unloading curves. It indicates the inability of the material to dissipate all the strain energy as heat, resulting in the deterioration of the product over time.

9. **Resilience** is the capacity of the material for strong strain energy in the elastic range. It is approximated as the area under the force-deformation curve up to the elastic limit point.

10. **Visco-Elasticity** is a combined liquid-like and solid-like behaviour of most agricultural products in which the stress-strain relationship is time-dependent. When the behaviour is such that the stress-strain ratio is independent of stress magnitude and only a function of time, the behaviour is said to be linear visco-elastic. However, most agricultural products behave in a non-visco-elastic manner in which the stress-strain ratio is dependent on both stress magnitude and time.

11. **Visco-Plasticity** occurs when the magnitude of the stress is such that the strain is non recoverable. The material becomes visco-plastic.

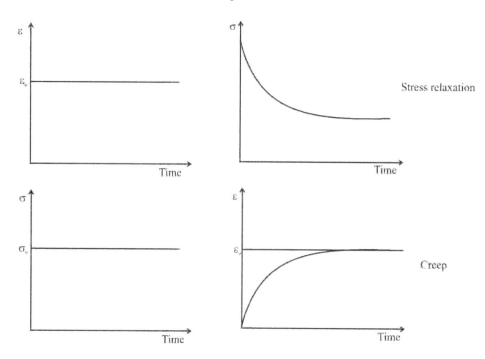

Fig. 2.8: Creep and Stress relaxation

32

12. **Stress Relaxation** is the decay of stress with time when the material is suddenly deformed and the strain held constant over time. The rate of stress decay or the time required for the stress to decay to a certain constant value is the relaxation time (Fig. 2.8)

13. **Creep** is the increase in deformation with time when the material is suddenly subjected to a constant stress (dead load). The increase is usually asymptotic (Fig. 2.8). The rate at which the retarded elastic deformation takes place in a material creeping under a dead load is the retardation time.

In their analysis of the mechanical properties of plantain fruit, Asoegwu, et al. (1998) found that the peel, cross-sectional diameter and loading rate of mature unripe plantain fruit play an important role in their mechanical behaviour. Bio-yield and rupture energies, stresses and strains were found to increase with an increase in loading rates for peeled plantain fruits. They also found that the modulus of deformation varied with the loading rate and cross-sectional area for both peeled and unpeeled unripe plantain fruits. The peel has been found to have a higher penetration resistance than the pulp (Asoegwu, 1991; Asoegwu, 1996).

CHAPTER 2 Review Questions

2.1 Give the significance of conveying machines in terms of how they are employed in modern plants for handling agricultural/biological products.

2.2 Enumerate all the ten characteristics of the transported materials to be considered when selecting the type and parameters of conveying machines.

2.3 Why can't efficiency, in all cases, characterize the mechanical properties of a conveying machine? Define the other parameter for this purpose.

2.4 In your own opinion, what do you consider as the most important attributes in mechanical conveying in the agricultural industry?

2.5 Why is the containerization of agricultural products important in conveying fruits and vegetables?

2.6 Define bulk density, particle density, void ratio, bulk porosity, surface area and specific surface. What is their importance in the design of materials handling equipment?

2.7 What is the difference between free-flowing and non-free-flowing materials? How do they relate to angles of repose and surcharge? What parameters influence the angle of repose?

2.8 What equations describe the volume and surface area of prolate spheroids, oblate spheroids, frustum of right cones?

2.9 Define particle diameter, volume diameter, surface diameter, surface mean diameter, volume mean diameter of agricultural products, and state the equations for their derivation.

2.10 Define sphericity, roundness, specific surface, equivalent diameter, shape coefficient, volume surface mean diameter, particle hardness, rigidity, abrasiveness of agricultural materials.

2.11 What parameters constitute the aerodynamic properties of agricultural products? Define terminal/critical velocity.

2.12 Give the equation of the resultant force acting on an agricultural product in a vertical stream of air. What is the condition of steady state and the equation of the critical velocity?

2.13 Define stress and strain and enumerate their types.

2.14 Sketch a force - deformation curve for an agricultural material and define the characteristic points of the curve.

2.15 Define toughness, elasticity, plasticity, visco-elasticity, stress relaxation, and creep.

3

CLASSIFICATION OF CONVEYORS

3.1   INTRODUCTION

The first concern in classifying, studying or selecting some materials handling components or systems is the material involved which could be liquid, free-flowing or non-free flowing materials in that order of convenience of handling.

However, by changing the physical form of the solid, as pelleting hay or grinding corn cobs, a non-free flowing material can be made relatively free-flowing. Or by suspending solids in water or any other fluid, either non-free flowing or free flowing material can be treated as a liquid.

3.2   CLASSIFICATION BY FEATURES:

Conveyors are usually classified by their features (i.e. their distinct or outstanding part or quality).

1.   **By the method of transmission of the driving force to the load.**
    i.   Mechanically driven (from electric, hydraulic or pneumatic drive).
    ii.   Gravity conveyors (moved by product own weight)
    iii.   Hydraulic and pneumatic conveyors (by airflow or water jet).
    iv.   Electrodynamic induction pumps.

2.   **By the kind of application of the driving force (this determines the design of conveying machines).**
    i.   Traction types with a traction element for the transfer of the driving force. Also, the loads move together with the traction element, e.g. belt, chain, bucket, escalators, elevators. The traction element may be belt or chain or cable or push bar etc.
    ii.   Traction-less type without the traction element. Here, loads are translated by rotational or oscillatory movement of the working elements of the conveying machine e.g. screw, gravity, vibrating and oscillating, roller and rotating, pneumatic and hydraulic conveyors and handling systems (Alexandrov, 1981).

3.   **By the kinds of load transferred.**
    (i) bulk load or (ii) unit load (or piece load).

4. **By the direction and path of moving the loads, which could be straight or curved (Fig. 3.1).**

    (a) Vertically closed machines convey loads in a vertical plane along paths consisting of a single rectilinear section (horizontal, vertical or inclined) or a combination of rectilinear sections (horizontal + inclined, horizontal + vertical, etc.).

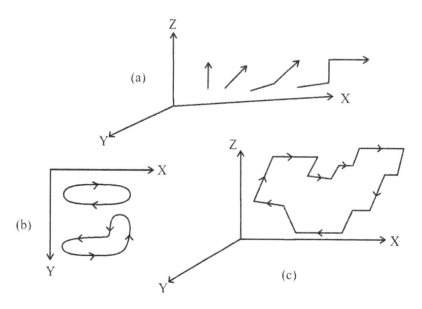

Fig. 3.1: Paths of moving the load by conveying machines
(a) Vertically; (b) Horizontally and (c) Spatial

    (b) Horizontally closed machines operate in one horizontal plane at the same level along a closed path of a particular shape.

    (c) Spatial (tri-dimensional) machines are placed in space and have a path consisting of horizontal, vertical and inclined sections. This group of machines are becoming very popular in modern industries.

5. **By the nature of motion of the load-carrying element.**
    Conveying machines may be of the continuous or intermittent motion types whose load-carrying elements may perform lateral, reciprocating, rotational or oscillating movements. Continuous motion may be by endless or self-crossing mechanisms and could be cyclic. Intermittent motion is a stop-and-go affair without any particular path or course. Reciprocating motion is a to-and-fro movement over the same fixed path of travel.

6. **By the course followed.**
This indicates the freedom of movement such as a fixed route (as with an industrial railroad) or a variable route (as with two-wheel hand trucks or forklifts which use any route the operator elects).

7. **By application and models.**
This indicates how the handling machines are applied and which models are there. They are divided into stationary, moving, re-adjustable, portable and moveable.

3.3 CLASSIFICATION BY FUNCTIONS:

How a conveying machine carries its load is also a characteristic distinct from the elements of motion it is capable of imparting. Some support their loads in platforms, trucks and belt conveyors, while others carry their loads suspended below the actuating machine as in cranes and monorails. There are also self-loading machines which are capable of picking up, transporting and setting down their loads without additional handling as in low and high lift platforms and fork trucks. Material handling equipment are mechanisms which incorporate certain elements into handling functions for easy identification and classification. Their functions are grouped as follow (Table 3.1):

Table 3.1: Classification of Conveyors by function

FHE	How load is carried		Path Travelled						Course followed		Kind of motion		
			In profile				On plane						
	Supported	Suspended	Horizontal	Vertical	Inclined	Declined	Straight	Curved	Fixed	Variable	Continuous	Intermittent	Reciprocating
T_p	√		√				√	√	√	√		√	
E_l	√	√		√					√		√		√
C_o	√	√	√		√	√	√	√	√		√		
T_f		√	√	√	√	√	√	√	√	√		√	
S_l	√	√	√	√			√	√		√		√	

FHE	-	Functional Handling Equipment

T_p — Transporting, E_l — Elevating

C_o — Conveying, T_f — Transferring

S_l — Self-loading

i.   **Transporting:** uses equipment equipped with running gears for moving loads over normally horizontal, straight or curved paths along fixed or variable routes.

ii.   **Elevating:** includes mechanisms for raising and lowering materials vertically with continuous, intermittent or reciprocating motion.

iii.   **Conveying:** defines gravity and power-driven machines for moving materials over horizontal, declined or inclined paths which though fixed, may be straight or curved with continuous and sometimes reciprocating motion.

iv.   **Transferring**: is a group of machines which lift their loads, move them through the air and deposit them along a fixed path or in an area defined by the limits of the machine itself, the motion being intermittent.

From the above analysis, it is obvious that the principle of operation of all lifting and conveying machines are:

(a)   **Intermittent motion**: whereby the machines are periodically loaded and unloaded when the machines are stopped.
The operating cycle includes
    i.   Stopping and gripping or lifting the load
    ii.   Motion with the load
    iii.   Stopping for unloading and
    iv.   Return motion without the load.

Examples are cranes, lifts, trucks, loaders, aerial train ways and cable ways, scrappers etc.

(b)   **Continuous motion**: which is characterized by non-stop motion of bulk or unit loads along a given path. The working motion (with the load) and the return motion (without the load) of the load-carrying elements of the machine occur simultaneously. There is no stopping for loading and unloading. The principal purpose of continuous conveying machines is to transport loads along a particular path, distribute loads among a number of destination points, deliver them to stores, transfer products from one technological operation to another, and ensure a desired pace of a production process.

3.4   SELECTION OF CONVEYING MACHINES:

Many factors are important in the accurate selection of a conveyor system. It is important to know how the conveyor system will be used beforehand. Some individual areas that are helpful to consider are the required conveyor operations, such as transportation, accumulation and sorting, the material sizes, weights and shapes and where the loading and pickup points need to be. Proper selection of conveyors makes it possible to integrate component parts into a smooth, efficient and functional materials handling system and greatly depends upon the nature of application and on the type of material to be conveyed (Olanrewaju et al., 2017).
:

Because there are many conveying machines of different sizes and capacities in the market, it becomes imperative to make informed choice for the solution of a particular conveying job. It is important to consider the job from a systems approach point of view. A systems approach to a bulk material conveying problem requires two major steps:

i.   Thorough analysis of the conveying job, including the overall objectives of the entire facility, and

ii.   The evaluation and complete investigation of all available methods of accomplishing the objective and final selection of equipment that will do the job most effectively.

These ensure that

-    The required quantities are conveyed in the time available
-    The desirable characteristics of the materials are maintained or established and flexibility is provided
-    The system is economically sound

Some questions that may lead to the appropriate engineering and selection of suitable conveyors are as follows:

Why   - the purpose of handling and processing equipment
What - the material and its handling characteristic
When - operating cycles in hours/day and days/week
Where - location and layout
How much - capacity, flow rate and power requirements
How Strong - structural requirements of stability
How expensive - cost effectiveness
How long - life expectancy, wear.

In analysing the conveying job to be done, the following steps are taken.

a.   Define and establish conveying objectives, volume of material to be conveyed to what distance, material processing methods and future requirements.

b.   Take and note data on the material (lump size, weight, etc.) that would influence choice of methods of conveying.

c.   Take and note plant space limitations likely to affect size or type of conveying equipment

d.   Review the existing conveying equipment (for continued suitability and efficiency before selecting a new one).

In evaluating and selecting a new conveying equipment,

i.   Note types of suitable conveying equipment

ii.   Compare these types as to efficiency, adaptability, operating and maintenance costs.

iii. Then, make final selection.

The **basic design principles** and data to be considered before selection of conveyors are:

- Select the conveyor according to the characteristics of the conveyed material e.g. bulk density, moisture content, angle of repose, etc.
- Ensure stability of conveyor under all normal working and climatic conditions both inside and outside.
- Make weight of conveyor low compared with the weight of the transported product.
- Consider the required conveying speed as well as the starting and run-out time.
- Insist that the throughput capacity corresponds with that of the following system.
- Do not exceed rated capacity and speed; define the motor power with safety factor to take up several full-load starts.
- Minimize conveying length by connecting subsequent operations with one another.
- Consult the constructor of the conveyor when making alterations such as increasing flow or changing unloading or loading points.
- Standardization on conveyors and their components will increase the speed of repair and maintenance and reduce stalling of the operation.
- Avoid spillage and environmental (noise and dust) pollution.
- Include return brakes to all vertical and inclined conveyors to prevent backsliding of the carrying element.
- Use gravity whenever and wherever possible.
- Ensure smooth changes of the product flow direction by careful design of conveyor transfer points to avoid product build-up, dust generation etc.
- Include the working intensity and conditions in the design *ab initio*. (Hours/day and days/week).

The above steps show that a strong engineering preference should be evaluated quantitatively by agricultural engineers in terms of estimated increased life, reduced operating delays and lower maintenance cost.

So, the principal criteria for the selection of a conveying machine, are:

a. It should satisfy the whole complex of basic engineering requirements (based on thorough engineering design), yet
b. Its application should be efficient and provide the most economical selection on a first cost or operating cost basis (in order that the unit cost of conveyance will be low).

The above conditions must ensure reliability of the system and satisfy labour protection and safety engineering requirements.

3.4.1 Engineering Responsibility

The agricultural engineer in charge of the materials handling section of a processing plant (i.e. in the Buyer's Engineering Department) will:

i. Completely define the conveying problem (including preliminary arrangement drawings).
ii. Completely specify the machine in terms of capacity, power, dimensions etc.
iii. Evaluate all the bids from an engineering point of view.
iv. Be responsible for all the engineering works of coordination, installation, the tune-up of the equipment and the training of operating personnel.

In performing the above functions, the agricultural engineer is not finding a solution but finding the best possible solution to the problem by determining both the methods of conveying and the equipment to be used in a given material handling situation, having analysed the options available. The engineer must study the manufacturing, distribution and capacity requirements as well as the product quality requirements so that after handling the product, there should be no change in the properties of the product. He can then recommend the best equipment for handling the material.

In studying the above requirements, the engineer's responsibilities are:

i. That the quality of the material is retained after handling
ii. That the capacity handled must equal the capacity specified, e.g. for a livestock feed that requires only 100 kg of corn/h from the silo to the mill, a 99 or 101 kg/h delivery will not give the specified capacity.
iii. That in recommending an equipment that should deliver 50 tons/h at normal conditions, consideration should be given to the loading, unloading, starting and stopping rates. Otherwise, the equipment will not deliver at that rate because of cleaning-up time, not working at full capacity at start and stop operations etc.
iv. That the best possible equipment is selected so that the feeding rate gives the specified capacity.

3.4.2 Engineering Factors

The following engineering factors must be considered when selecting a material handling equipment.

i. The physical and mechanical characteristics of the load material in terms of size requirements of its preservation, the reliability of the machine operation (moist or sticky materials), and the conditions of labour protection (dusty materials).
ii. The throughput capacity of the machine in terms of machine speed (increase in machine speed decreases the quantity of the load per unit length of the conveyor).

iii.   The direction, length and configuration of the load transfer path. Machines are designed for an optimal length and optimal path and selection should be of machines that will perform the task in a single flight in order to minimize reloading.

iv.   The methods of loading and reloading. Self-loading and self-unloading machines are preferable to minimize power consumption.

v.   There should be assured good compatibility of production processes with the transfer path of the machine.

vi.   There should be data on the production and climatic conditions of the environment in which the machine must operate. These affect the servicing and lubrication of mechanisms and may be prone to explosion- or fire-hazard.

3.4.3   Economic Factors

The business of conveying in processing plants is to maximize profit and make processes efficient. The following economic indices are used to compare various versions of a machine before selection based on technical and economic analysis.

i   Capital expenditure (initial cost) for purchasing and installation.

ii.   Maintenance and operating costs. Maintenance includes day-to-day care and routine replacement of worn-out components; operating cost includes power.

iii.   The number and productivity of workers engaged in loading and unloading operations.

iv.   The pay-back period for capital expenditure.

The annual economic effect of application of a machine is the resulting saving (economy) in all production resources (labour, materials, capital costs, etc.) which can be obtained in the running of the machine, given as Eqn. 3.1.

$$E = (C_b - C_n)A \qquad\qquad ...3.1$$

where E = annual economic effect, C_b and C_n are recalculated expenditures per unit of production or work performed on base (old) equipment and on recommended (new) equipment respectively; and A = annual output of production. The C = recalculated expenditure is given by Eqn. 3.2.

$$C = C_u + E_r K \qquad\qquad ...3.2$$

where C_u = cost/unit of production, E_r = rated coefficient of effective investment (E_r = 0.15), and K = unit capital investment in production.

The machine with the highest annual economic effect when compared with other versions is considered optimal and selected.

An analysis of an annual economic effect contains:

i.   Results of the calculation of economic indices
ii.   Findings and recommendations
iii. Addenda containing documents in support of initial data.

It gives the engineering characteristics of versions, compares their principal parameters, life times, etc. The results contain:

i.   Service life (including probable obsolescent).
ii.   Production cost and the price of the machine
iii. Coefficient of equivalence between base and new versions.
iv.   Unit capital investment in production
v.   Accompanying capital expenditures of the consumer for a machine.
vi.   Annual output of the new machine
vii. Total volume of production.

As a result of the above evaluation, the most economical selection of the conveying equipment can be made. The specific type of conveyor can be determined and its size, speed and power requirements obtained. The conveyor must also be found to be reliable.

3.4.4   Reliability of Conveyors or Conveyor Systems

Evaluation of reliability, availability and maintainability (RAM) plays an important role in performance estimation of conveying systems (Ahmadi et al., 2019). The reliability of a machine or machine system is its capacity for performing the specified functions and retaining for a long time the specified operating characteristics (speeds, throughput capacity, power consumption, etc.) invariable within given limits according to the conditions of operation, maintenance and repair. It comprises of:

i.   **No failure operation** - the ability of the machine to operate uninterruptedly without failure for a definite time.
ii.   **Durability** - the property of the machine to remain operable (intact) up to an ultimate state as established by the schedule of repairs and maintenance.
iii. **Maintainability** - the quality of a machine that helps prevent and detect failures in its operation and eliminates their consequences by maintenance and repair.

Because a conveyor is an inseparable part of a whole production system; whereas flow-line production is based on conveyor transfer of articles; and a stop of a conveyor system interrupts the manufacturing process and decreases the production output, the reliability of the conveyor is very important. The principal complex characteristic of reliability is the availability factor (K_a) given as Eqn. 3.3.

$$K_a = T_f / (T_f + T_r) \qquad\qquad \dots 3.3$$

where T_f = life time to failure (h) and T_r = mean time of restoration which characterizes the maintainability of a machine (h).

High values of K_a can be attained by improving the design of conveyors or by increasing the labour input for maintenance and repairs.

3.4.5   Common Types of Conveyors

The following types of conveyors are used extensively in agricultural processing industries: belt, chain, screw, bucket, vibrating and oscillating, pneumatic and hydraulic, gravity and roller, fans and blowers, etc. (Bhandari, 2003). Cranes, trucks, carts, forklifts might be considered as intermittent conveyors and are not discussed in this Book.

3.5   SAFETY PRECAUTIONS AND OPERATION RULE:

Conveyors can be unsafe if installed or operated improperly and they do not conform to specified safety regulations and codes. These potential safety hazards apply to conveyor maintenance as well as conveyor operation (Thayer, 2017). Conveyors must be designed, constructed and used with the idea of safety in mind. Consideration of safety must be paramount in the design and selection of conveyors. Note is taken of the following points:

- All shearing and squeezing points as well as all rotating drive components should be guarded by expanded metal or steel bolted guards.
- All movable covers should be designed not to open while the conveyor is working and the conveyor should not be started while the covers are open.
- Inspection opening should be guarded by covers
- Manual stop-lock devices should be provided and their resetting done by an authorized person only after the fault has been rectified.
- Audible or visual alarm signals should be provided
- Equipment should be easily cleaned and lubrication points easily accessible without removing the guards.
- Sharp edges and corners as well as projecting and protruding parts should be avoided.

The above points will make the operation of and on conveyors safe.

Care and Maintenance of Conveyor Systems

A conveyor system is often the lifeline to a company's ability to effectively move its product in a timely fashion within the plant. The steps that a company can take to ensure that it performs at peak capacity, include regular inspections of sensitive parts, close monitoring of motors and reducers, keeping key parts of the conveyor in stock, and proper training of maintenance personnel and staff that work with the conveyors.

By obeying the operation rule, many failures of components and equipment-related accidents can be prevented. Some of these rules are:

- Use conveyors for the duties they were designed, they should not handle products not originally specified.
- No alterations should be made on the conveyor without the consent of the designer and constructor.
- Always use gangways, staircases, steps, ladders or platforms and do not step on, or ride on or even cross-over a moving or operable conveyor.
- You must be qualified and competent to operate a conveyor; no unauthorized person should operate or interfere with it.
- Stopping devices should be easily accessible and checked periodically for proper functioning.
- Do not restart a stopping device unless after you have determined the cause of the stoppage and repaired the fault.
- Conveyors not having their safety devices and guards in good working order are unsafe for operation and should not be operated until adequate safety is ensured.
- Repair and removal of safety devices and protective enclosures or panels can only be carried out by competent personnel after stopping the conveyor and putting the restart device inoperable.

Conveyors have been the most important transport media in transferring products from one point to another in the industry. The monitoring and protection of these conveyors are very important as the occurrence of faults may affect the whole operation with negative economic effects. The protection of the conveyors which was carried out using Relay Logic methods proved to have several disadvantages. A new method now focuses on the monitoring, controlling and protecting the conveyors from various types of faults occurring in them using programmable logic controller (PLC) (Vanitha, 2019). Four important types of faults that occur frequently in conveyors, such as belt sway fault, pull chord fault, zero speed fault and fire protection are sensed and rectified by programmable logic controller which has a high degree of safety, accuracy and easy to maintain and monitor.

CHAPTER 3 Review Questions

3.1   How are conveyors classified by features and by functions?

3.2   What are the principles of operation and the operating cycles of all lifting and conveying machines?

3.3   What steps are required in a systems approach to solving a bulk material conveying problem?

3.4   In analysing and evaluating a conveying job to be done, what steps must be followed?

3.5   What are the principal criteria for the selection of a conveying machine?

3.6   What is the responsibility of an engineer in charge of a material handling section of a plant?

3.7   What engineering factors are considered when selecting a material handling equipment?

3.8   What economic indices are used before the selection of conveyors?

3.9   Categorize conveyors into traction and traction-less machines. What are the major differences between them?

3.10   What is reliability and what does it comprise? What is the availability factor of conveyors?

3.11   What basic design principles and data are to be considered before the selection of conveyors?

3.12   What engineering questions lead to the selection of suitable conveyors?

3.13   Why is safety important in selecting conveyors? Name some safety points to be considered in an agricultural processing plant.

4

■■■■■■■■■■■■■■■■■■■■■■■■■

GENERAL THEORY OF CONVEYORS

4.1 INTRODUCTION

Conveying machinery is intended to move loads in a continuous stream without stops for loading and unloading. The device may be horizontal, inclined or vertical and is used to convey the materials in a path predetermined by the design and having points of loading and discharge fixed or selective. Unit loads are moved piecemeal, while bulk materials are moved in a continuous or intermittent flow. However, continuous flow is preferable.

The bulk weight of conveyed material is influenced by its physical properties of moisture content and size of its constituent particles and lumps. Increase of these properties increases the bulk weight which will also increase the energy requirement of the conveying machine. Bulk density may be used to differentiate loads into categories (Table 4.1).

Table 4.1 Categorization of conveyed materials by bulk density

LOADS	BULK DENSITY (t/m³)
Light (sawdust, powders, flour, peat, coke etc.)	< 0.6
Medium (grains, coal, slag, seeds, nuts)	0.6 to 1.1
Heavy (sand, gravel,)	1.1 to 2.0
Extra heavy (ore, pebbles, stone)	>2.0

Most agricultural products have medium bulk density at 14% m.c. (wet basis). Their densities increase with increase in moisture content.

The uniformity of particles or lumps in a conveyed load influences the process of conveying. Thus, loads are usually graded (sized) or ungraded (un-sized). Particle-size distribution is done by sieving. The relationship between size of largest lumps (a_{max}) and the size of the smallest lumps (a_{min}) determine whether the conveyed load is graded or ungraded. When $a_{max}/a_{min} > 2.5$ the load is ungraded; and if $a_{max}/a_{min} < 2.5$ the load is graded. Graded load is characterized by the mean size of lumps: $a_{mean} = (a_{max} + a_{min})/2$

Also, the free-flowing characteristics of small grains and pellets permit their handling by a number of different devices. In many agricultural handling and processing systems the flow plan requires both vertical and horizontal movement of the material. Some devices, such as the screw or auger can be used at any elevating angle. Others are best suited for vertical or horizontal operation only. Still, other devices such as the chain and flight conveyor are practical for angles from horizontal to a maximum elevating angle of about 45°.

4.2   THROUGHOUT CAPACITY

The throughput capacity is an important engineering parameter to be determined and designed for when selecting any materials handling equipment. It is the main criterion of conveyors performance and defined as mass or volume of the load moved per unit time. There are two main types of throughput capacity:

i.   **Operating (actual)** throughput capacity (Q_{ac} or V_{ac}) generally applied to a complete system or plant, takes care of down-time periods for maintenance and repair and idle time to move, relocate or adjust the equipment. It is expressed as tonne/year, per day or per hour.

ii.   **The rated (design)** throughput capacity (Q or V_s) is the quantity of goods that can be conveyed per unit time at the ideal (as found in design) operating conditions (no equipment breakdowns) with complete filling of the load carrying element of the machine sustained for specific time periods as required by the operation.

While the rated (design) throughput capacity depends on complete filling of the load-carrying element operating at constant rated speed, the operating (actual) throughput capacity depends on the degree of filling of the load-carrying element and on the utilization of the machine in time. Their relationship is given in Eqn. 4.1.

$$Q = \varrho V_s \qquad \qquad \text{...4.1}$$

where Q = capacity by weight (t/h); V_s = **volumetric throughput capacity** (m³/h); ϱ = bulk weight (t/m³)

The total coefficient of utilization (K_{ac}) of the machine used in conveying materials in-plant is given as Eqn. 4.2.

$$Q_{ac}/Q = V_{ac}/V_s = K_{ac} \qquad \qquad \text{...4.2}$$

$K_{ac} = K_n \ x \ K_t \ x \ K_a$ where K_{ac} = total coefficient of utilization of the machine; K_n= coefficient of non-uniform loading; K_t= coefficient of machine utilization in time, and K_a = coefficient of availability; Q_{ac} & V_{ac}= actual capacity; Q and, V_s = design (rated) throughput capacity.

When moving bulk materials in a steady stream, the design (rated) capacity is given as Eqn. 4.3.

$$V_s = 3600AV \ (m^3/h) \text{ and } Q = 3600\varrho AV \ (t/h) \qquad \qquad \text{...4.3}$$

where   A = cross-sectional area of the material (m^2); V = conveying speed (m/s).

The **rated (design) throughput capacity** is the principal characteristic that determines the main design parameters of a machine (the dimensions of the load-carrying element and its operating speed) at which this capacity can be attained.

The cross-sectional area, A, of a bulk material heaped on a stationary surface varies with the properties of the material and is determined by the angle of repose θ_r ie by the steepest angle to the horizontal at which the heap will stand. When the conveyor is moving on a horizontal path, the angle of repose reduces to the angle of surcharge as in Eqn. 4.4.

$$\theta_s = (0.35 - 0.70)\theta_r \qquad \qquad ...4.4$$

In whatever way the transportation is done whether in continuous flow, portion wise or piecemeal, the principal parameters which determine throughput capacity are the average quantity of goods per unit length of the load-carrying element and the operating speed of transportation. Throughput capacity is given as Eqn. 4.5.

$$V_s = 3.6q_v V \ (m^3/h); \ Q = 3.6q_v \varrho V \ (t/h); \ Q = 3.6q'v \ (t/h) \quad ...4.5$$

where ϱ = density of bulk load (t/m^3); V = speed of conveyor (m/s); q' = linear density of load (kg/m); q_v = volume conveyed per unit length of conveyor (1/m); V_s = volumetric throughput capacity (m^3/h).

If the load-carrying element is in the form of a trough or pipe of cross-sectional area, A_o and a filling coefficient,ψ, the cross-sectional area of the load in the trough $A = A_o\psi$, the throughput components are given in Eqn. 4.6.

$$q_v = 1000A_o\psi; \ q' = 1000A_o\varrho\psi; \ V_s = 3600A_o V\psi; \ Q = 3600A_o V\varrho\psi \quad ...4.6$$

Using buckets and like containers of volume, i_o (1iter) with filling coefficient, ψ, the volume of load in a bucket, $i = i_o\psi$, and if the spacing between the buckets = t_b (m), the volume of load per metre of conveyor length (l/m) will be given as Eqn. 4.7.

$$q_v = i/t_b = i_o\psi/t_b \qquad \qquad ...4.7$$

and the hourly throughput capacity (m^3/h or t/h) is as in Eqn. 4.8.

$$V_s = 3.6iV/t_b = 3.6i_o\psi V/t_b ; \ \ Q = 3.6iV\varrho/t_b = 3.6i_o\psi V\varrho/t_b \qquad ...4.8$$

For unit loads of average mass m (kg) conveyed individually or in groups of z pieces, with spacing of t_a (m), then the mass of load per unit length of conveyor (kg/m) is Eqn. 4.9.

$$q' = m/t_a = mz/t_a \qquad \qquad ...4.9$$

and the throughput capacity (t/h) will be Eqn. 4.10.

$$Q = 3.6mV/t_a = 3.6mzV/t_a \qquad \qquad ...4.10$$

If t = time interval between pieces or portions of load as in Eqn. 4.11.

$$t = t_a/V \qquad \qquad ...4.11$$

and the hourly throughput capacity (pieces/h) as in Eqn. 4.12.

$$Z = 3600/t = 3600V/t_a = 3600z/t = 3600zV/t_a \qquad \qquad ...4.12$$

The mass capacity (t/h) is given in Eqns. 4.13 and 4.14.

$$Q = 3.6q'V \qquad \qquad ...4.13$$

$$Q = mZ/1000 \qquad \qquad ...4.14$$

For an inclined conveyor, the inclination depends on the angle of repose in motion, i.e. the surcharge angle, (θs) and the nature of the load and to a lesser extent to the sag of the belt between idler rolls. The throughput capacity of an inclined conveyor is given by Eqn. 4.15.

$$Q_{inc} = 3600K_\beta AV\varrho \qquad \qquad ...4.15$$

where K_β = slope factor or flowability factor.

Also, since the throughput capacity for a bulk load in continuous flow is a function of the cross-sectional area of the load-carrying element or linear load and the operating speed, one can determine the geometrical parameters of the load-carrying element like the cross-sectional area of the trough or pipe, the width of the belt, the shape of the trough, etc. to ensure required throughput capacity.

4.3   DRIVE MOTOR POWER:
Most conveyors are powered by one or more drive motors whose capacity must be great enough to overcome the traction resistance required to convey the load in the system. Movements of loads could be vertical, horizontal or both. When a load of throughput capacity Q (t/h) is moved

vertically to height H (m), the power to drive the motor required for lifting the load alone (without resistance i.e. useful power, kW) will be Eqn. 4.16.

$$N_u = 1000QHg/3600 \times 1000 = QH/367 \qquad \dots 4.16$$

Taking g = 9.81m/s^2

The required power (kW) of the motor is Eqn. 4.17.

$$N = Nu/\eta = QH/367\eta \qquad \dots 4.17$$

where η = efficiency of the conveying machine (0.75 - 0.90). For horizontal conveying of loads, H = η = 0.

However, a parameter known as coefficient of resistance of motion defined as the ratio of the resistance forces developed in motion of a load to the gravity force of that load is used to characterize the mechanical properties of the horizontal conveying machine.

If q' (kg/m) = mass of load/metre length of conveying machine; L (m) = the length of the conveying machine; µ' = total coefficient of resistance; then the force of resistance to friction is given by Eqn. 4.18.

$$F_{fr} = q'gL\mu' \simeq 10q'L\mu'(N) \qquad \dots 4.18$$

(Taking g = 10m/s^2)

The power to overcome this force of resistance (kW) is given by Eqn. 4.19.

$$N_{fr} = F_{fr}V/1000 \cong 9.81q'L\mu V/1000 \cong QL\mu'/367 \qquad \dots 4.19$$

The total power (kW) to perform the lifting work and the work of overcoming the horizontal friction forces will be given as in Eqn. 4.20 or Eqn. 4.21.

$$N = N_u + N_{fr} = (Q/367)(H + L\mu') \qquad \dots 4.20$$
Or
$$N = (V\varrho/367)(H + L\mu') \qquad \dots 4.21$$

The relationship between the coefficient of resistance µ' and efficiency η can be determined by equating Eq. 4.17 and Eq. 4.20 as in Eqn. 4.22.

$$QH/367\eta = (Q/367)(H + L\mu') \qquad \dots 4.22$$

To give Eqn. 4.23.

$$\mu' = (H/L)\left(\frac{1}{\eta} - 1\right) \qquad \ldots 4.23$$

Two conditions can be deduced from Eq. 4.23. For horizontal motion, H = 0; vertical motion, H = L and

$$\mu' = (1/\eta) - 1 \qquad \ldots 4.24$$

CHAPTER 4 Review Questions

4.1 Define throughput capacity of a conveying system, the types of this capacity and the parameters that they depend on.

4.2 Define coefficient of resistance and state its relationship with efficiency. State when it comes into use.

4.3 What is the importance of bulk density in calculating throughput capacity?

4.4 Throughput capacity is a function of two parameters. What are these two parameters, which when chosen determines the geometrical parameters of the load-carrying element? Give examples of these geometrical parameters.

4.5 If a load is moved horizontally and vertically through length (L m) and height (H m) and the throughput capacity is Q (t/h) and the total coefficient of resistance is μ', calculate the total power (kW) required to perform the lifting work and to overcome friction forces. If the motor efficiency is η, find the general relationship between μ' and η. What is this relationship for horizontal motion and vertical motion?

4.6 When calculating the throughput capacity of continuous conveying machines, name three cases of transportation usually considered and the two expressions of throughput capacity used. Define and give the unit of each symbol used in the expression.

4.7 What factors influence the bulk weight of conveyed materials? How does bulk weight affect the power requirement of conveyors?

5

BELT CONVEYORS

5.1 INTRODUCTION

The belt conveyor is the most widely used conveyor capable of carrying a greater diversity of products at higher rates and over longer distances than any other kind of continuously - operating mechanical conveyors (Wang et al. 2010). It is an endless belt operating between two or more pulleys with the belt and its load supported at intervals on idlers. Belt conveyors are being employed to form the most important parts of material handling systems because of their high efficiency of transportation. It is significant to reduce the energy consumption or energy cost of material handling sector (Memane and Biradar, 2015). It transports materials horizontally or on an incline, either up or down. The drive pulley and the discharge chute are at the head end (Fig. 5.1) and there is usually a take-up pulley which is preferably located just behind the head pulley. The feed chute and loading skirts are located near the tail pulley. Even though it is preferable to have the discharge chute at the head pulley, discharge may take place at any point along the carrying run of the belt by the use of a movable diagonal scraper or diverter or tripper or by tilting of one or more of the idler pulleys.

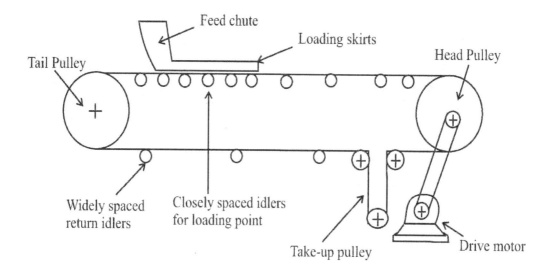

Fig. 5.1: Belt conveyor arrangement

Whereas the belt conveyor is capable of handling practically all pulverized, granular and lumpy materials ranging from flour, cement, sand to fruits and coal, there are some limitations to its applicability. These include:

1. Temperature of the environment or the transported material must not be too high to scorch the belt which may be unprotected canvas.

2. Angle of repose and angle of friction of the materials determine the slope of the belt. The slope must not be too steep to cause the transported materials to slip upon themselves or on the belt, and fall back along the slope.

3. The centre-to-centre distance of the belt should be within the body strength of the belt so that it could withstand the expected load and operating tension, and if possible, the belt should have a resistant surface.

At the discharge or transfer point, the conveyed material must be properly directed onto another conveyor, into some sort of processing equipment, or into storage. The design of the transfer points which is very problematic is influenced by the material characteristics, the speed of the conveyors, the relative direction of the conveyors, and the rate of handling. Added to the simple over-the-head-pulley-discharge are other discharge systems that make use of plough scrappers and trippers. For the conveyor to function effectively, it must be loaded centrally at a uniform, steady rate. Also, this system has some advantages of low cost, stable performance and obvious effect in energy saving (Dong and Luo, 2011).

5.2 CHARACTERISTICS OF THE BELT CONVEYOR

The characteristics of the belt conveyor that make it very suitable in many agricultural processing operations include:

1. The carrying capacity can be high (130 - 342 m^3/h) (Sahara and Kuroda, 1985). This is because it can operate satisfactorily at high belt speeds (1.0 - 300.0 m/min) (Comley, 1982); it could be fitted with sidewalls or transverse slats or cleats or the surface textured so that it can operate on a steep incline.

2. Materials can be conveyed over long distances (> 80 m) (Roberts and Hayes, 1979), but there is a limit to the angle of elevation (< 30°).

3. With all loads carried on antifriction bearings, mechanical efficiency is high (> 85%).

4. With proper design, maintenance is low and infrequent, service life is long but initial investment is high. Whenever, there is need for maintenance, it entails a lot of expenditure.

5. Since there is no relative motion between the material and the belt and no friction damage on the material, the belt conveyor does not impart any mechanical damage on the material.

6. In the design of belt conveyor, vibration and impact in the belt, are ignored but the static design is into account.

7. In order to ensure the safety of the operation of the conveyor, the safety factor of its design is Increased (Li et al., 2016).

5.3   ESSENTIAL ELEMENTS OF BELT CONVEYORS

1.   **The belt** consists of the carcass or core and the cover which protect the carcass. The belt forms the moving and supporting surface which carries the transported material along. It is the main element and represents a large proportion of the capital cost of the belt conveyor system. The belt carcass is usually composed of two to ten plies or layers of woven fabric bonded together with rubber. It provides the tensile strength to transmit power to move the load, absorbs the energy at loading points, provides the rigidity to support the conveyed material, and the flexibility required for the trough and wrapping around drums and pulleys. The choice of the belt cover is influenced by the nature and quantity of the material conveyed, the way the material is fed to the belt, the belt speed and the environment under which it is operating. To make the belt endless, vulcanized splice and mechanical fasteners are used.

2.   **The idlers**, which may be flat or made into troughs, support the carrying and return strands of the belt. Return idlers are flat while carrying idlers may be made into troughs to increase the carrying capacity of the belt. They must be properly spaced to avoid spillage, eliminate sagging and minimize wear of the belt. Idlers determine, to a great extent, the performance, power requirements, and operating cost of a belt conveyor. Their choice takes into account belt width and speed, nature of load and size of particles. Trough idlers (Fig. 5.2) are between two to five (usually three) rollers, 100 mm to 175 mm diameter with roll angle of between 30 – 35° depending on the angle of repose of the conveyed material. With deep troughs this roll angle can reach about 70° to the horizontal.

To assist in belt alignment, idler rolls are made with slight forward tilt angle of not more than 3° between the axis of the wing rollers and that of the central one in the direction of belt travel. The carrying idler spacing is a function of belt width and belt tension and must take into account the bulk density of the conveyed material. They are closely spaced at the feed-end or tail end and widely spaced at the head-end or discharge-end of the conveyor.

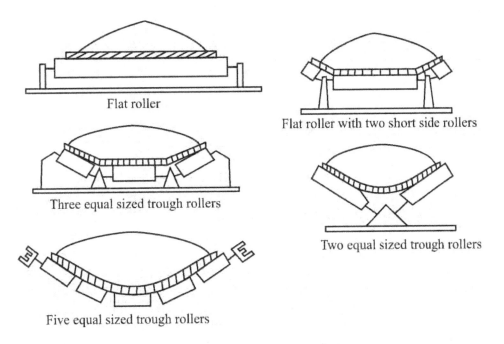

Flat roller

Flat roller with two short side rollers

Three equal sized trough rollers

Two equal sized trough rollers

Five equal sized trough rollers

Fig. 5.2: Trough idlers

Return idlers which are usually in contact with the top cover of the belt are made of various forms of rubber disc or spiral wire rollers in order to guard against the build-up of fine materials on them. Transition idlers that transform a trough belt to a flat belt should be properly selected and positioned to eliminate excessive stretching of the belt which may result in permanent belt damage, wear and spillage. All in all, idler selection is governed by type of service, belt speed and characteristics of the conveyed material (weight and lump size).

3.   **The drive** may be at either end of the conveyor, but preferably, at the head-end where tension is minimal (Fig. 5.3a). A conventional belt drive is usually used. The effectiveness of the drive depends upon the difference in tension between the tight side and the slack side of the belt, the friction between the belt and the drum, and wrap angle of the belt to the drum. The drive power can be increased by increasing either the coefficient of friction (using rubber lagging) or the wrap angle (using an idler pulley) (Fig. 5.3b) or by providing a multiple drive (Fig. 5.3c). It is important that the inherent tension in the belt needed to maintain the drive is provided by either a screwed tension device or a gravity take-up or by hydraulically or electrically powered automatic take-ups. The take-ups stretch and eliminate belt sag, prevents belt slips and maintains traction between the belt and drive pulley. The electric motor provides the power which is transmitted via belt, gears or fluid and eddy-current couplings.

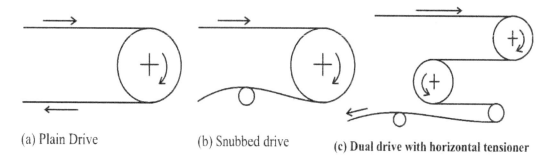

(a) Plain Drive   (b) Snubbed drive   **(c) Dual drive with horizontal tensioner**

Fig. 5.3: Types of drives

4.   Materials are loaded on to the belt conveyor by hand, from a **hopper or bunker or funnel** with a gate valve or from a proceeding feeder which may be an apron, screw, belt or vibratory conveyor. The most important thing is that, whichever method of loading that is used, it must provide continuous steady flow of the material which must be distributed uniformly about the centreline of the belt. This requires some sort of transfer chute or skirt whose length is about two or three times the belt width from the feed point in the direction of belt travel. Discharge of materials from the belt conveyor usually takes place at the head-end drum. However, if discharge is required before the head-end drum, a plough scrapper or a tripper may be used.

5.   **The power unit** of a belt conveyor comprises the driving motor and the power transmission. The electric motor is commonly used with some kind of speed reduction mechanism which employs either worm gears or a gear train of straight-cut or helical gears. Power transmission is taken from a motor/gearbox unit with torque control devices (fluid or eddy-current couplings) which change the torque or speed characteristics of the electric motor to suit the conveyor and allow for flexibility in the selection of the motor. The fluid couplings are preferred to eddy-current couplings which are bulky and expensive.

6.   For bulk products that tend to stick to the surface of the belt, **belt cleaners** are placed at head end to minimize the build-up of material on snub pulleys and return idlers which can drastically shorten the life of the belt and interfere with conveyor performance. These cleaners may be rotary rubber brushes of rotating, helical and vibrating scraper blades of steel or rubber material. Other suitable cleaning devices of mechanical, shaking and rinsing nature exist. The most popular being the mechanical whose edge is made of rubber, plastic, wood or steel pressed against the belt by a spring, weight or a hydraulic cylinder at a pressure of 0.10-0.15 kg/cm^2 of scrapper width. The coefficient of friction of the scrapper on the belt is 0.60-0.75.

5.4   BELT CONVEYOR DESIGN CONSIDERATIONS/CALCULATIONS

In designing a belt conveyor, i.e. selecting the standard size of the belt and calculating the required power of the conveyor drive, the following **design considerations** must be carefully studied to provide design data base for the development of a reliable and efficient belt conveyor system that will reduce cost and enhance productivity while simultaneously reducing dangers to workers operating them (Gupta and Dave, 2015).

a.   The physical characteristics of the material to be conveyed (kind, bulk density, lump size distribution, coefficient of friction, angles of repose and surcharge, height to which it can be piled on the belt, etc.)

b.   The average and maximum rated load-carrying capacity which depends on the speed and width of the belt as well as the cross-sectional area and density of the load on the belt.

c.   Data on the expected operating conditions of the load which include moisture content, dustiness, cohesiveness, abrasiveness and corrosiveness, hot or cold, oily or smooth.

d.   Methods of loading and unloading

e.   The diagram of the designed belt conveyor with dimensions

f.   Arrangement and operation of the belt conveyor.

In actual design we rely on empirical tables and charts for the preliminary analysis. However, in using trough belts for agricultural materials, we determine the type and size of the belt, the speed of the belt required to deliver the required amount of material and the power of the motor that moves the conveyor.

5.4.1 Belt Width Determination

Belts have been standardized (DIN 22107). Different sizes are available in the market. The belt size selected must be the next larger one to the calculated width.

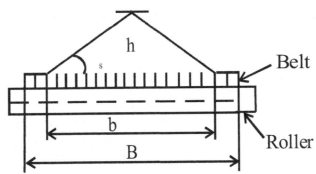

Fig. 5.4: Loaded belt on straight roller

For belts supported by straight rollers (Fig.5.4) the load forms an equilateral triangle whose area is given in Eqn. 5.1.

59

$$A = bhK_p/2 = 0.25b^2K_p tan\theta_s \quad (m^2) \qquad \ldots 5.1$$

where b = (0.9B - 0.05) is the carrying width of belt (m) and B = belt width (m),
      h = $0.5b\ Tan\ \theta_s$ is the height of the load layer on the belt (m),
     K_β= coefficient of reduction of load layer cross-section on an inclined portion of conveyor
        (slope factor)
     θ_s = $(0.35 - 0.70)\ \theta_r$ = angle of surcharge (deg.). (Table 5.1)
     θ_r= angle of repose (deg.).
     θ_s = $1.11\theta_r - (0.1\ \beta + 18°)$
where β = trough idler angle (deg.).

Table 5.1   **Values of Coefficient K$_\beta$**

Mobility of particles	θ_s (deg.)	Angle of incline of conveyor (deg.)				
		1 – 5	6 - 10	11 - 15	16-20	21-24
High	10	0.95	0.90	0.85	0.80	-
Medium	15	1	0.97	0.95	0.90	0.85
Low	20	1	0.98	0.97	0.95	0.90

    If we assume that the surface of the conveyed material is parabolic (Fig. 5.5), the cross-sectional area A of the load stream on a flat belt is given in Eqn. 5.2.

$$A = (b^2 tan\theta_s)/6 \quad (m^2) \qquad \ldots 5.2$$

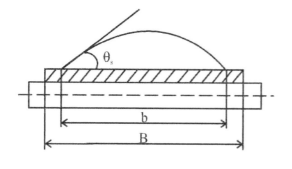

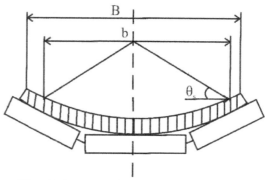

Fig. 5.5: Load of parabolic surface   ***Fig. 5.6: Trough carrying load***

Table 5.2   **Values of the shape factor U**

Type of Idler		Mobility of material particles

	Angle of incline of side rollers	High	Medium	Low
Straight	**0**	158	240	328
Trough (with 3 rollers)	20 30 45 60	393 480 580 582	470 550 633 620	550 625 692 662

For trough belts (Fig. 5.6) one may express the cross-section (A) of the load stream in terms of the contact perimeter, b, of the material on the belt using a shape factor U (Table 5.2), as Eqn. 5.3.

$$A = Ub^2 \qquad (m^2) \qquad\qquad ...5.3$$

For flat belt, $U = (Tan\ \theta_s)/6$ = a function of the transverse profile and the surcharge angle.

Again, for trough belts, the total cross-sectional area is the sum of the upper equilateral triangle and the lower trapezium whose sides are formed by the side roller supports.

Allowance is made for edge clearance (distances between conveyed material and edge of the belt). The minimum belt width B_{min} in terms of the contact perimeter b as given in BS.5934/ISO 5048 is Eqn. 5.4.

$$B_{min} = 1.11b + 0.056 \qquad (m) \qquad\qquad ...5.4$$

Based upon a uniform feed to the conveyor, the cross-sectional area of the load on the belt determines the belt conveyor capacity for a certain belt speed. The carrying capacity of the inclined belt conveyor Q_{max} is therefore given by Eqn. 5.5.

$$Q_{max} = 3600AV\varrho K_\beta = UV\varrho K_\beta b^2 \qquad\qquad ...5.5$$

where A = cross-sectional area of load on belt (m²); V = speed of belt (m/s); ϱ = density of conveyed material (t/m³); $U = 3600\ A/b^2$ = coefficient of the cross-sectional area of load on belt (a shape factor); b = contact perimeter of load on belt (m).

For a given shape factor and belt speed, the required minimum width of belt (m) required to transport material of known bulk density at a given rate is found from $B_{min} = 1.11b + 0.056$ to be Eqn. 5.6.

$$B_{min} = 1.11(0.05 + \sqrt{Q_{max}/UV\varrho K_\beta} \qquad \dots 5.6$$

Belt width is also influenced by lump size of the conveyed material which has a bearing on the surcharge angle. For surcharge angles of 20°, allow maximum lump size of one-fifth of the belt width, i.e. B/5; up to 30° surcharge angle, allow maximum lump size of one-tenth of the belt width i.e. B/10; for less than 10° surcharge angle, allow maximum lump size of one-third of the belt width i.e. B/3. The minimum belt width is given in Eqn. 5.7.

$$B_{min} = Xa + 200 \qquad (m) \qquad \dots 5.7$$

where X = coefficient of lump size (3.5 for graded, 2.5 for ungraded); a = maximum linear size of lump (m).

Final belt width is chosen from normal series: 300, 400, 500, 600, 650, 800, 1000, and 1200, 1400, 1600, 1800, 2000, 2500, and 3000 mm. It is essential that belts have edges from 50 to 100 mm wide free from the load. Also, the belt speeds must conform to the belt width and type of material transported (Table 5.3).

5.4.2 Belt Speed Determination

In choosing the belt speed, the following factors are considered - nature of material (lump size, abrasiveness, dustiness), belt width, carrying capacity required and the belt tensions (i.e. the way the belt is loaded and unloaded), and the application and location of the conveyor. The speed of the conveyor belt is selected from Table 5.3. Modern normal series conveyor belt speeds frequently employed are 0.8, 1.0, 1.25, 1.6, 2.0, 2.5, 3.15, 3.5, 4.0, 5.0, 6.3, 8.0 and 10.0 m/s. If care is taken over the dynamic design of the system, belt speeds can be above 15 m/s (Harrison and Roberts, 1983).

5.4.3 Appropriate calculation of belt tension and drive motor power.

In belt drive, the force or power is transmitted through friction between the drum or pulley and the belt. There exists a difference of tension in the belt on the two sides of the drive pulley (Fig 5.7.). Power is transmitted by means of this difference in the tension in the belt running on the pulley (T_1) and off the pulley (T_2). Both are related by Eqn. 5.8.

$$T_1 > T_2 \qquad (N) \qquad \dots 5.8$$

And Euler's equation is Eqn. 5.9.

Table 5.3 **Recommended speeds of conveyor belts (m/s) unloaded through drive pulley (without intermediate unloading)**

Transported material	Belt width (mm)				
	400-500	650-800	1000-1200	1400-1600	2000-2500
Non and low-abrasive, abrasive powdered and granular (salt, sand, coal, peat)	1.1-1.60	1.75-2.5	2.5-4.0	2.75-4.0	3.5-5.0
Abrasive, fine and medium sized lumpy materials (gravel, rock, ore, slag, ash)	1.0-1.60	1.65-2.0	2.0-2.5	2.5-3.15	3.5
Abrasive (very), large size lumpy material (stone, rock, ore, coke, coal)	1.1-1.50	1.50-1.75	1.6-2.0	2.0-2.5	2.5-3.15
Fine, dusty and dry powdered dusting material	0.8-1.0	0.8-1.0	0.8-1.0	0.8-1.0	0.8-1.0
Brittle lumpy materials	1.25	1.6	1.6	2.0	2.0
Grain heavy (rye, wheat)	1.5-2.0	2.25-3.0	2.5-4.0	3.0-4.0	3.0-4.0
Light Oats	1.5-2.0	2.25-3.0	2.5-4.0	2.5-3.0	2.5-3.0
Fruits and vegetables	0.8	0.8	1.0	1.0	1.0

$$T_1 = T_2 e^{\mu\alpha} \qquad (N) \qquad\qquad \ldots 5.9$$

where μ = coefficient of friction between belt and pulley (0.10-0.5) and
  α = wrap angle on belt pulley (radians).

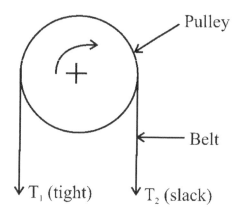

Fig. 5.7: Belt tensions

For proper functioning of the belt, two conditions must be fulfilled: -

(i)   Power is transmitted from drum to belt without slippage

(ii)   Excessive sag does not occur between any pair of idlers

The effective tension $T_e = (T_1 - T_2)$ and the wrap or drive factor C_w are two important parameters in conveyor belt design given as Eqns. 5.10 and 5.11.

$$T_1/T_2 = e^{\mu\alpha} = (1 + C_w)/C_w \qquad \text{...5.10}$$

also, $$C_w = T_2/T_e = T_2/(T_1 - T_2) \qquad \text{... 5.11}$$

The circumferential traction force that can be transmitted to a pulley neglecting losses on the pulley owing to belt rigidity is given by Eqn. 5.12 with the drive factor Eqn. 5.13.

$$T_e = T_1 - T_2 = T_2(e^{\mu\alpha} - 1) = T_1(e^{\mu\alpha} - 1)/e^{\mu\alpha} \qquad \text{...5.12}$$

$$\therefore C_w = 1/(e^{\mu\alpha} - 1) = 1/(C_p - 1) \qquad \text{... 5.13}$$

where C_p = pull factor which determines the traction capacity of the driving pulley.

The wrap or drive factor, C_w depends upon coefficient of friction and angle of contact. Tables 5.4 and 5.5 show the values of the two parameters. From Table 5.5, $\mu = 0.25$ was used for bare pulleys and $\mu = 0.35$ for lagged pulleys. When a screw tension device is used, the value C_w is increased by 20% if a bare drum is used and 30% if the drum is lagged, in order to guarantee sufficient slack side tension T_2.

Table 5.4: Coefficient of friction (μ) values

Belt Condition	Bare Drum/Pulley	Lagged Drum/Pulley
Dry	0.3 - 0.5	0.35 - 0.6
Light wet	0.2	0.25 - 0.4
Clean wet	0.1	0.2 - 0.3
Wet and dirty	0.05 - 0.1	0.2

Table 5.5: Wrap or drive factor (Cw)

Type of Drive	Angle of Contact $\alpha°$	Gravity Take-up		Screw Take-up	
		Bare	Lagged	Bare	Lagged
Plain	180	0.84	0.50	1.2	0.8
Snubbed	200	0.72	0.42	1.0	0.7
	210	0.66	0.38	1.0	0.7
	220	0.62	0.35	0.9	0.6
	240	0.54	0.30	0.8	0.6
Dual or Tandem	380	0.23	0.11	0.5	0.3
	420	0.18	0.08	-	-

In order to ensure the required margin of friction force on a driving pulley, a somewhat increased tension of the running-off section of the belt is maintained to prevent sagging. It is given by Eqn. 5.14.

$$T_2 = K_a T_e / (e^{\mu\alpha} - 1) \qquad (N) \qquad\qquad\qquad …5.14$$

The rated tension of the running-on belt section that ensures no slippage is given by Eqn. 5.15.

$$T_1 = T_2 e^{\mu\alpha} = K_a T_e e^{\mu\alpha} / (e^{\mu\alpha} - 1) \qquad (N) \qquad\qquad …5.15$$

where K_a = 1.1 - 1.2 is the margin of adhesion between the belt and drum, T_e = effective pull.

The power of the drive motor is given as Eqn. 5.16.

$$N = K_a T_e V / 1000\eta \qquad (kW) \qquad\qquad\qquad …5.16$$

where V = speed of conveyor (m/s); η = total efficiency of the drive mechanism (0.8-0.9).

5.4.4 Belt Sag

Minimum belt tension must be sufficient to prevent excessive belt sagging between the idlers. It is customary to limit the belt sag to a minimum of 3% of the span between the carrying idlers. Different methods of estimating belt sag and sag tension have been developed (Roberts and Hayes, 1979; Spivakosky and Dyachkov, 1985; Colijn, 1985; MHEA, 1986; Woodcock and Manson, 1987). The admissible belt sag is given by Eqn.5.17.

$$Max\ sag = WL^2/(8T_{min}) = (0.025 - 0.03)L \qquad (m) \qquad \qquad ...5.17$$

where W = weight of belt and load/metre length of belt; L = idler spacing and T_{min} = minimum belt tension. In the carrying run, the belt tension should not be less than the value enabling the belt sag to be kept at not greater than 2.5 - 3.0%

5.4.5 Idler Spacing

The best idler spacing depends on the weight of belt and load, belts sag, type of idler and its loading rate, width and stiffness of belt and tension within the belt. Carrying idlers are placed between 0.93m and 1.6m apart (Table 5.6) while return idlers are universally placed 3m apart.

Table 5.6: Suggested Normal Spacing for Carrying Idlers

Belt width (mm)	Pitch of carrying idler sets (mm)		
	Bulk density of conveyed material (kg/m³)		
	400-1200	1200-2000	2000-2800
400 - 650	1650	1500	1425
650 - 900	1500	1350	1275
900 - 1050	1350	1200	1125
1050 - 2000	1200	1000	925

5.4.6 Binder Belts

These contact the conveyor belts at the drive pulley and increase the grip of the belt on the pulley by increasing the driving traction in the belt by ΔT (Eqn. 5.18) without changing the tension in the belt.

$$\Delta T = T_b\ (e^{\mu\alpha} - 1)/e^{\mu\alpha} \qquad (m) \qquad \qquad ...5.18$$

where T_b = the tension in the binder belt, thus making effective tension as in Eqn. 5.19.

$$T_e = T_1 - T_2 = (T_2 + T_b)(e^{\mu\alpha} - 1) \qquad (N) \qquad \qquad ...5.19$$

5.4.7   Power Requirements of Belt Conveyor

The power requirement of a belt conveyor is determined by the following factors:

i   Coefficient of friction between the drive pulley and the belt.
ii   The tension in the belt.
iii   The arc of contact between the pulley and the belt; as shown in the proceeding sections.

It consists of three components all of which include friction factors for idler rotation and belt/load flexing: -

a   Power to move belt system horizontally, with or without load, overcoming empty or loaded friction
b   Power to move belt system vertically (raise or lower load)
c   Power to overcome friction from ancillary components (e.g. skirts, scrappers, trippers, etc.)

·   When belt with capacity Q (t/h) lifts or lowers the load to or from a height H(m), the required power in kW is given by Eqn. 5.20.

$$HP_{vert} = \pm\ QH/367 \qquad (kW) \qquad\qquad (kW) \qquad\qquad\qquad ...5.20$$

If horizontally, the power requirement will depend on the frictional resistance to the movement of the belt, which varies directly with the capacity Q and the belt length L(m).

This horizontal power is given by Eqn. 5.21.

$$HP_{hori} = (C.f.L/367)(3.67G_mV + Q) \quad (kW) \qquad\qquad ...5.21$$

where C = friction factor depending on belt length = Put $C = (L + L_o)/L$; f = friction value for idler; L = length of belt (m); G_m= total belt weight plus equivalent weight of rotating idler rolls (kg/m of conveyor); V = belt speed (m/s); L_o = equivalent length factor (m) (See Table 5.7)

From Eqn. 5.20 and Eqn.5.21 which include the power to initiate motion from rest and power to move belt and load, standards such as DIN # 22101 and Goodyear (1982) have the following belt horsepower (kW) Eqns. 5.22 and 5.23, respectively.

For DIN // 22102
$$HP = HP_{vert} + HP_{hori} = (C.f.L/367)(3.67G_mV) \pm QH/367 \ (kW) \quad ...5.22$$

Table 5.7: Length Factor (L₀) and Composite Friction Factor (C₀)

Class of Conveyor	Friction Factor C_o	Length Factor L_o
Permanent, well-aligned conveyors with normal maintenance	0.022	61
Temporary, portable or poorly aligned conveyors which may or may not operate at below - 5oC.	0.03	46
Conveyors requiring restraint of the belt when loaded.	0.012	145

For Goodyear (1982)

$$HP = f[(L + L_o)\,C.f.\,L/367^2](3.67 G_m V) \pm QH/367 \quad (kW) \quad \ldots 5.23$$

The effective pull (T_e) in kg is given by Eqn. 5.24.

$$T_e = 102 HP/V \quad (kg) \qquad \ldots 5.24$$

from which the values of T_1 and T_2 can be got using Euler's equation: - From Eqn. 5.23, Eqns. 5.25 and 5.26 are got.

$$T_e = C_o(L + L_o)(G_m + [Q/3.67V] + [QH/3.67V])(kg) \qquad \ldots 5.25$$

$$HP = T_e V/102 \quad (kW) \qquad \ldots 5.26$$

where $C_o = C \times f$ composite friction factor (Table 5.7).

5.4.8 Determination of Drive Pulley Diameter

The standard size of pulley and reducer is chosen from the rated torque of the drive shaft given by Eqn. 5.27.

$$M = 0.5 \, K \, T_e \, D_p \quad (Nm) \qquad \ldots 5.27$$

where K = 1.1 - 1.2 = coefficient for unaccounted losses; D_p = diameter of driving pulley (m).

Conveyors with rubber-fabric belts have pulley diameters given by Eqn. 5.28.

68

$$D = K_{pl} \; x \; K_p x \, i \; = \; D_p K_p \qquad \text{(m)} \qquad \qquad \text{...5.28}$$

where K_{pl} = coefficient on ply fabric strength ranging from 125 - 200 m/ply pieces depending   on ply fabric strength ranging from 55 to 400 N/mm for cotton, capron and amide;
K_p = coefficient depending on type of pulley (Table 5.8); i = no. of plies on belt carcass.

Table 5.8: Coefficient K_p depending on pulley type

Pulley Type	K_p
Single	1.0
Double and take up	1.1
Belt tension <= 60% of allowable	0.8-0.85
Belt tension > 60% of allowable	0.90
Deflector pulleys	0.50

Using Eq. 5.28, there are standard series of pulley diameters: 160, 200, 250, 315, 400, 500, 630, 800, 1000, 1250, and 1400, 1600, 2000, and 2500 mm. Table 5.9 gives the recommended minimum drive pulley diameters. Tail and take-up pulley minimum diameters are the same as given in the Table (5.9) for 40 - 60% classification. Also, slack slide bend pulleys with belt wrap less than 150mm may have minimum diameters 150mm less than the diameters recommended.

In choosing the pulley, the loads to be sustained must be known and these are known from designing the shafts for strength and the bearings for life expectancy. Shaft diameters are chosen according to the type of duty the belt will do (Table 5.10).

The life of the bearing under a certain set of condition is given by Eqn. 5.29.

$$L = (C/L)^b f_l \qquad \qquad \text{...5.29}$$

where L = life in millions of revolutions; C = dynamic load rating (kg); P =actual radial load (kg);
b = 3.0 for ball bearing and 3.3, for roller bearing; f_l = lubrication factor.

To convert to hours, use Eqn. 5.30.

$$L = (1,000,000/60n)(C/L)^b f_l \qquad \text{(hr)} \qquad \qquad \text{...5.30}$$

where n = rotational speed in rpm.

Table 5.9   Recommended minimum drive pulley diameters (mm)

No. of piles	*RMA 35 Fabric			RMA 43 and 50 Fabric			RMA 60 and 70 Fabric		
	% normal belt tensions			% normal belt tensions			% normal belt tensions		
	over 80-100	over 60-80	over 40-60	over 80-100	over 60-80	over 40-60	over 80-100	over 60-80	over 40-60
3	457	356	305	508	457	356	610	508	406
4	508	457	406	610	508	457	762	610	508
5	610	508	457	762	610	508	914	762	610
6	762	610	508	914	762	610	1067	914	762
7	914	762	610	1067	914	762	1219	1067	914
8	1067	914	762	1219	1067	914	1524	1219	1067
9	1219	1067	914	1372	1067	914	1676	1372	12191
10	1372	1219	1067	1524	1219	1067	1829	1829	372

* RMA = Rubber Manufacturers Association.

Table 5.10: Belt Duty and Shaft Diameter

CEMA SERIAL NUMBER	BELT DUTY	ROLL DIAMETER (mm)	Shaft Diameter (mm)
A	Light	100 - 125	16
B		100 - 125	17 – 19
C	Medium	100 - 125	19
D		150	19 – 25
E	Heavy	150 - 175	30 – 32

5.4.9 Belt Conveyor Frictional Resistance

The total resistance to be overcome which determines the working pull transmitted at the drive pulley is the sum of all the frictional resistance throughout the length of the conveyor plus the resistance due to lifting the load.

The power developed by the belt conveyor is required to overcome frictional effects in the system and for increasing the load's potential energy.

These frictional resistances are as follows: -

a   **Belt Friction Resistance (F_1):** is as a result of the movement of the belt over the other moving masses: idlers, pulleys, drums, etc. of the conveyor (Eqn. 5.31).

$$F_1 = \mu_1 M_c g L \quad (N) \qquad \qquad \qquad ...5.31$$

where μ_1 = belt friction factor ranging from 0.018 to 0.06 depending on various service conditions; M_c = mass of all moving or rotating parts per unit length of the belt (kg/m).

b   **Load Friction Resistance (F_2):** is concerned with the horizontal movement of the conveyed material involving the total mass of the material on the belt. (Eqns. 5.32 and 5.33).

$$F_2 = \mu_2 M_l g L \quad (N) \qquad \qquad \qquad ...5.32$$

where $M_l = Q_{max}/V$ = mass of load per unit length of belt (kg/m); L = conveying distance (m); V = speed of belt (m/s).

$$\therefore F_2 = \mu_2 (Q_{max}/V) g L \qquad \qquad \qquad ...5.33$$

For design purposes, $\mu_1 = \mu_2 = 0.025$, thus making the main resistance F_m to be Eqn. 5.34.

$$F_m = F_1 + F_2 = 0.025 g L [(M_c + Q_{max})/V] \quad (N) \qquad ...5.34$$

c   **Load Slope Resistance (F_3):** arising from the gravity force of the load being conveyed upwards or downwards (Eqn.5.35).

$$F_m = F_1 + F_2 = 0.025 g L [(M_c + Q_{max})/V] \quad (N) \qquad ...5.35$$

where H = change in vertical elevation; (+) upwards and (-) downwards.

d   **Load Acceleration Resistance (F4):** arises if the load being fed on to the belt has an initial velocity V_o in the direction of the belt whose velocity is V (Eqn. 3.56).

$$F_4 = M_l V(V - V_o) = Q_{max}(V - V_o) \qquad (N) \qquad\qquad ...5.36$$

This most important secondary resistance when combined with the other secondary resistance gives F_n which for conveyor belts longer than 80 m is Eqn. 5.37.

$$F_n = \mu_3 F_m \qquad (N) \qquad\qquad ...5.37$$

where μ_3 = secondary resistance coefficient (Table 5.11); and F_m = main resistance (Eq. 5.34)

Table 5.11: Secondary Resistance Coefficient μ_3

Conveyor Length (m)	68	70	100	200	500	1000	2000	5000
μ_3	1	0.98	0.77	0.46	0.2	0.098	0.05	0.025

From the above analysis, the total resistance F_r (N) in steady motion of the belt will equal the required driving traction force of the motor i.e. the effective tension of the belt (Eqn. 5.38).

$$F_r = F_1 + F_2 + F_3 + F_n = (1 + \mu_3)F_m + F_3 = T_e \qquad\qquad ...5.38$$

T_1 and T_2 can then be estimated using Euler's formula. The motor horsepower (kW) is given by Eqn. 5.39.

$$HP_m = T_e V / \eta \qquad (kW) \qquad\qquad ...5.39$$

where η = motor efficiency (0.85 - 0.95).

5.4.10   Resistance at Loading and Unloading Stations
Belt conveyors operate successfully when they are loaded and discharged properly. The resistance at the loading and discharging points must be taken care of in the design of the system.

a.   **Loading Station:** If the conveyed material is fed into the skirt (Fig. 5.8) continuously at velocity V_o while the belt is moving at velocity V, the load will accelerate at uniform rate within distance L until its speed becomes V. The **kinetic energy**, K.E. gained in the period, if the throughput capacity is Q t/h, will be Eqn. 5.40.

$$K.E. = Q((V^2 - V_o^2)/3.6 \ x \ 2) \qquad\qquad ...5.40$$

At time t, the belt travelled the distance S_b (Eqn. 5.41).

$$S_b = (V_o + [(V - V_o)/2])t \qquad \qquad \text{...5.41}$$

And the load slides a distance S_L (Eqn. 5.42).

$$S_L = [(V - V_o)/2]t \qquad \qquad \text{...5.42}$$

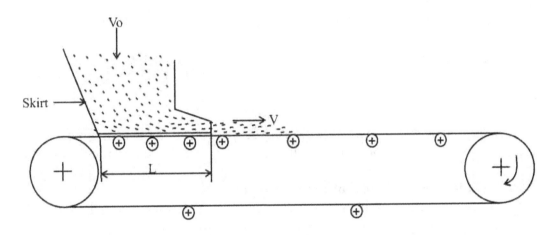

Fig. 5.8: Loading station of belt conveyor

If the constant friction exists between load and belt, the work, W, done by the sliding load will be less than the kinetic energy, K.E. of the load by a ratio of the sliding distance and the belt distance and given by Eqn. 5.43.

$$\therefore W = K.E.\,[\{(V - V_o)t\}/2\{V_o + (V - V_o/2)t\}] \qquad \text{...5.43}$$

The total energy due to the resting force, F_{ls}, at the loading station of the belt travelling at a speed of V m/s is given by Eqn. 5.44 and Eqn. 5.45.

$$F_{ls}V = K.E. + W = Q[\{(V^2 - V_o^2)V\}/3.6(V + V_o)] \quad (N.m/s) \qquad \text{...5.44}$$

$$F_{ls} = Q[(V - V_o)/(3.6)] \qquad \qquad \text{...5.45}$$

It will be observed that Eqn. 5.36 and Eqn. 5.45 are similar. However, allowing for friction of load against the hopper or skirt by introducing a factor $\Phi = 1.3 - 1.5$, the total resisting force at the loading station is given by Eqn. 5.46.

$$F_{ls} = \Phi Q[(V - V_o)/(3.6)] \qquad \text{(N)} \qquad \qquad \ldots 5.46$$

Note that inclined chutes help to minimize the kinetic effect of the load on the belt.

b. **Unloading Station**: If unloading is by ploughing, the frictional resistance force varies directly with the belt width and the load weight per unit length of the belt and is given by Eqn. 5.47.

$$F_{us} = (2.7 \text{ to } 3.6) \, q \, B \, (kg) \qquad \qquad \ldots 5.47$$

where 2.7 is for dusty material, 3.1 for granular material and 3.6 for small lumpy material; B = width of the belt; q = load weight per unit length of belt.

For a belt clearing device, the resisting force is given by Eqn. 5.48.

$$F_{cl} = \mu_4 B \qquad \text{(N)} \qquad \qquad \ldots 5.48$$

where μ_4 is the frictional coefficient of the clearing device in N/m of the belt width; $\mu_4 = 30$ to 50 for plough and $\mu_4 = (2 \text{ to } 6) \, V_b$ for rotary brushes having V_b as the circumferential speed.

5.4.11 Start-up and Braking Times of Belt Conveyors

The process of starting up the belt conveyor consists of starting it from the state of rest and accelerating all the moving masses to the rated conveyor speed. The maximum acceleration at which the load will still remain stable on the moving belt is given by Eqn. 5.49.

$$J_{max} = K \, g \, (f \, Cos \, \text{ß} - Sin \, \text{ß}) \qquad \text{(m/s}^2\text{)} \qquad \qquad \ldots 5.49$$

where K = 0.6 - 0.8 = safety coefficient; f = 0.8 - 1.0 = coefficient of friction between load and belt; and ß = angle of inclination of the conveyor. The acceleration is usually maintained within the range 0.1- 0.2 m/s^2 while the start-up time is determined by Eqn. 5.50.

$$t_{st} = I_c^2 ng/[375(M_t - M_{st})K_{el}] \qquad \qquad \ldots 5.50$$

where I_c^2 = flywheel moment of all moving parts (kg.m^2); n = rated rotational speed of motor (rpm); M_t = average starting torque of motor (Nm); M_{st}= moment of static resistance forces in steady motion of belt (Nm); K_{el} = coefficient of elastic elongation of belt which is 0.7 - 0.8 for rubber-fabric belts; 0.9 - 1.0 for rubber cable belts; $M_t = (1.2 - 1.5) \, M_{st}$ or taken from electric motor catalogues (Nm).

For inclined conveyors, the component of gravity force of the inclined sections at the highest load will be greater than or equal to the resistance to belt motion in all the conveyor sections. This condition requires that a braking torque must be produced at the drive pulley shaft to fully stop the conveyor at braking time given by Eqn. 5.51.

$$t_{br} = [I_c^2 ng/\{375(M_t - M_{st})K_{el}\}] \quad (s) \qquad \qquad ...5.51$$

and the designations are as in Eq. 5.50.

5.5   MAINTENANCE OF BELT CONVEYORS

Maintenance of a belt conveyor starts with good design and proper installation of carefully stored and transported belts which must be correctly spliced and fitted. Maintenance operations have to be carefully planned during the design phase, with the adoption of appropriate safety measures for the maintenance crew and others (Martinetti, 2017). On the other hand, preventative measures have to be put in action for maintenance operations and daily work. Maintenance during operation includes check-up of all mechanical and electrical components, the structures and access ways for unusual noise from idlers, shafts, bearings, drives, belts, cleaners and scrapers (Boumans, 1985).

In general, maintenance of a belt conveyor when in operation or at rest must check and rectify the following: -

1.   The electric motor for overheat, unusual noise, nonalignment with shaft and vibration. The bearings must be lubricated and the mounting bolts fastened. With the connection box closed, the motor fan should be free-moving and the air outlet and motor casing dust-free.

2.   The gear box for overheated casing, unusual noise, nonalignment with shaft and vibration. The oil level should be checked as well as leaking seals and the oil changed periodically. Fastening of mounting bolts may solve some earlier mentioned problems.

3.   The couplings for overheat and nonalignment with, and tightness on, the shaft, keys and key-ways. The shock-absorbers, oil level and oil leaks should be checked also in the case of hydro-dynamic coupling.

4.   The pulleys (drive, return and take-up) for wear, cracks, tightness, fatigue, caking of the pulley lagging, slip between belt and drive pulley and proper functioning of scrapers and cleaners. The pulley must be tight on the shaft and the bearings must be greased on schedule to reduce overheat and unusual noise. Proper alignment and easy shaft movement in the bearings must be tightened and securely fastened respectively.

5.   The trough and return idlers for free movement of rolls and even or excessive wear. Idler units square with the centerline of the conveyor will not permit grooves worn by rubbing belts.

The application of new developed solutions in belt conveyor designs leads, among others, to considerable reduction in operational costs of transport systems, while ensuring their high reliability and service life at the same time (Zheng and Hou, 2014; Mazurkiewicz, 2015). Nonetheless, there are still areas that pose challenge to both research and development. It is absolutely vital from a practical point of view that the service life of conveyor belt joints be made longer owing to the fact that these joints are crucial elements of the entire conveyor transport system (Mazurkiewicz, 2014).

5.6 SAMPLE PROBLEMS:

Q.5.1: Dry beans of density 0.87 t/m³ and medium mobility were conveyed on an inclined troughed belt conveyor moving at 1.3 m/s. Trough and inclined angles are 20°. Determine the contact perimeter, the minimum belt width and the shape factor, if the throughput capacity is 286 t/h. Also, what will be the pitch of the carrying idlers?

Data: $\varrho = 0.87\ t/m^3$; $V = 1.3\ t/m^3$; $Q_{max} = 286\ t/h$; $\theta = \beta = 20^o$; $medium\ mobility$

Solution: From Table 5.1, $K_\beta = 0.90$; from Table 5.2; U = 470 (shape factor). Substituting the above values in Eqn. 5.5.

$$Q_{max} = UV\varrho K_\beta b^2,\ \text{we have}$$

$$286 = 470\ \text{x}\ 1.3\ \text{x}\ 0.87\ \text{x}\ 0.9\ \text{x}\ b^2$$

Therefore, b = 0.7732 m = **773.2 mm**.
From Eq. 5.4, Belt width, $B_{min} = (1.11b + 0.056)\ m = 0.9142\ m = **914.2\ mm**$.
Choose belt 1000 mm wide. From Table 5.7, the pitch of the carrying idlers is **1350 mm**.

Q.5.2: A belt conveyor conveys 150 t/h of rice paddy at a speed of 2.0 m/s, a distance of 80 m and a height of 5.5 m for storage. If the specific friction factor = 0.030 and length factor = 46, what will be the motor power rating, the power required to drive the pulley shaft and the maximum effective pull of the belt. Assume weight of belt and all rotating idler rolls to be 18 kg/m.

Data: $Q_{max} = 150\ t/h$; $V = 2.0\ m/s$; $L = 80\ m$; $H = 5.5\ m$; $C_o = 0.030, Lo = 46$; $G_m = 18\ kg/m$.

Solution
    (i)   Motor power rating = horizontal power requirements.
        From Eqn. 5.21:

$$HP_{hor} = C_o[(L + L_o)/367](3.67G_mV + Q_{max})$$

$$\therefore HP_{hor} = 0.03[(80 + 46)/367](3.6x18x2 + 150) = **2.91\ kW**$$

    (ii)   Power to drive the pulley shaft = total power requirement.

$$HP_{total} = HP_{hor} + HP_{vert} = 2.91 + (QH/367) = 2.91 + (150x5.5/367) = **5.16\ kW**$$

    (iii)   Max. Effective pull

$$P_{max} = T_e = 102HP_l/V = 102x5.16/2 = \mathbf{263.16\ kN}$$

Q.5.3: At the loading station of a belt conveyor, wheat grain was loaded at a velocity 0.6 m/s while the belt speed is 1.5 m/s. For the 7 seconds it takes the two velocities to be equal, calculate: -

    i   Distance travelled by the belt
    ii   Distance slid over the belt by the wheat grain;
    iii   The K.E gained by the wheat
    iv   The work performed by the sliding load
    v   The resisting force.

If the wheat is lifted 13.5m through a horizontal distance of 150 m, find the power required at the drive pulley shaft and the motor power.

$$\text{Assume } Q_{max} = 150\ t/h; C_o = 0.022; L_o = 61; G_m = 25\ kg/m;$$
$$\Phi = 1.3\ and\ \eta = 0.85$$

Data: $V = 1.5\ m/s; V_o = 0.6\ m/s, H = 13.5\ m; L = 150\ m; t = 7\ s$

Solution
    (i)   From Eqn. 5.41; distance travelled by the belt is: -

$$S_b = (V_o + (V - V_o)/2)t = (0.6 + (1.5 - 0.6)/2)x7 = \mathbf{7.35\ m}$$

    (ii)   From Eqn. 5.42, distance slid over the belt is: -

$$S_l = (V - V_o)t/2 = ((1.5 - 0.6)/2)x7 = \mathbf{3.15\ m}$$

    (iii)   From Eqn. 5.40, the K.E gained by the load is: -

$$K.E. = Q\ (V^2 - V_o^2)/(3.6\ x\ 2) = 150\ (1.5^2 - 0.6^2)/(3.6\ x\ 2) = \mathbf{39.4\ J}$$

    (iv)   From Eqn. 5.43, work done by sliding load is: -

$$W = K.E.(V - V_o)/(V + V_o) = 39.4\ (1.5 - 0.6)/(1.5 + 0.6) = \mathbf{16.9\ J}$$

    (v)   From Eqn. 5.46, the total resisting force is: -

$$F_{ls} = \Phi Q(V - V_o)/3.6 = 1.3x150(1.5 - 0.6)g/3.6 = \mathbf{478.2\ N}$$

(vi)   Power required at the drive shaft is got by using Eqn. 5.23: -

$$HP_{total} = [C_o(L + L_o)/367](3.67G_mV + Q) + (QH/367)$$

$$\therefore HP_{total} = [0.022(150 + 61)/367](3.67x25 + 150) + 150x13.5/367$$

$$= 3.64 + 5.52 = \textbf{9.16 kW}$$

(vii)   Motor power, HP_m =

$$HP_m = HP_{total}/\eta = 9.16/0.85 = \textbf{10.8 kW}$$

Q.5.4: Wheat of density 0.77 t/m³ and angle of repose 30° is to be moved 125 m horizontally by a flat belt conveyor at a rate of 250 t/h with the load cross-section area maintained at 0.03 m².
   a.   specify the belt width, surcharge angle, and belt speed.
   b.   how much power is absorbed in accelerating the grain, if the grain is moving horizontally at 1.45 m/s when it hits the belt?

Data: $V_o = 1.45\,m/s$; $L = 125m$; $Q_{max} = 250\,t/h$; $\varrho = 0.77\,t/m^3$; $\theta_r = 30^o$; and $A = 0.03\,m^2$

Solution:

a.   Put $\theta_r = 30^o$; $Surcharge\ angle = \theta_s = 1.11\theta_r - 18 = 1.11x30 - 18 = 15.3^o$

$$U = tan\theta_s/6 = \textbf{0.0456}$$

From Eqn. 5.5,

$$Q_{max} = 3600AVK_\beta\varrho; but\ K_\beta = 1\ for\ flat\ belts$$

$$\therefore V = Q_{max}/3600\varrho A = 250/3600x0.77x0.03 = \textbf{3.00 m/s}$$

From Eq. 5.2, $A = (1/6)b^2K_\beta tan\theta_s \quad but\ K_\beta = 1$

$$b = \sqrt{0.03x6/tan\ 15.3} = \textbf{0.8112 m}; \quad B_{min} = (b + 0.05)/0.9$$

$$= 1.11b + 0.056 = \textbf{0.9564 m}$$

Choose belt of width **1000 mm**

b.   The power absorbed in accelerating the load equals the product of the load acceleration resistance and belt speed.

From Eqn. 5.36, $F_4 = Q_{max}(V - V_o) = 250(3.00 - 1.45)/367 = \mathbf{1.056\ kN}$

Power for accelerating the grain is given by: $HP_{grain} = 1.056x3.00 = \mathbf{3.17\ kW}$

CHAPTER 5 Review Questions

5.1   Enumerate the advantages of a belt conveyor.

5.2   To what uses are belt conveyors put in an agricultural engineering enterprise? What factors influence the design of the transfer points in belt conveyors?

5.3   Name two public places where double-belt conveyors may be used.

5.4   Which three properties of the transported materials are most important in the design of belt conveyor and why?

5.5   Enumerate the basic components of a belt conveyor. What are the functions of the belt?

5.6   What initial data would you require for selecting a standard size belt and power of the conveyor drive? What factors influence the choice of the belt cover?

5.7   What does the factor C_o in the power equation of belt conveyor mean?

5.8   Define angles of repose and surcharge and the relationship between them.

5.9   What factors affect the cross-sectional area of a material on the belt conveyor? What factors limit the applicability of the belt conveyor?

5.10   What are the requirements of the conveyor belt? What factors affect the effectiveness and power of the drive of the belt conveyor?

5.11   What are the limitations of a belt conveyor?

5.12   What two conditions must be fulfilled for the proper functioning of the belt? What do you do to maintain the belt conveyor and why the maintenance?

5.13   Grains of a repose angle of $25°$ are moved by 60 cm belts on straight rollers. Calculate the carrying width of the belt, the height of the load layer and the cross-sectional area of the load. If the grain above has density 0.85 t/m^3 and the belt moves at 3.2 m/s, what is the carrying capacity of the conveyor? [b = 0.49; θ_s = 9.75°; h = 0.042 m; A = 0.0103 m^2]. *(Q_{mex} = 100.86 t/h).*

5.14   The off-tension of a belt conveyor is 1500 N. If the coefficient of friction is 0.35 and the wrap angle is $180°$, find the motor horsepower required to drive a belt conveyor at a speed of 3.5 m/s? Assume K_a = 1.15 and η = 0.85. *(N = 3.42 kW).*

5.15   If the belt length = 10 m and weight of belt and load per metre length of belt = 1.15 kg/m, find the maximum belt sag for a minimum tension of 500 N. *(Sag = 0.282 m).*

5.16   A 12 m belt conveyor travelling at 4.0 m/s has a total weight/metre of conveyor to be 1.45 kg/m. If friction is 0.35 and throughput capacity is 100 t/h, what is the horizontal horsepower required for this well-aligned conveyor? *(N = 8.44 kW).*

5.17   Calculate the total frictional forces to be overcome and the horsepower required by a $10°$ inclined belt conveyor of length = 70 m, mass of all moving parts = 0.60 kg/m, moving at 3.8 m/s upwards and conveying wheat at 150 t/h. The grain is fed at 2.0 m/s. Assume η = 0.85; μ_1 = μ_2 = 0.025 *(F_1 = 10.3 N; F_2 = 188.2 N; F_3 = 1308 N; F_4 = 75 N; F_r = 1581.5 N; HP = 7.07 kW).*

5.18   A belt conveyor is to be used to convey 50 t/h of rice paddies at a speed of 2 m/s a distance of 15 m and to a height of 6 m for processing. Calculate the motor power rating, the power required to drive the pulley shaft, the max effective pull, and the max and min tensions in the belt. Assume α = $180°$; f = 0.35; C_o = 1.30.

(HP$_{hor}$ = 2.66 kW; HP = HP$_{hor}$ + HP$_{vert}$ = 3.47 kW; T$_e$ = 177.2 N; T$_{max}$ = 265.7 N and T$_{min}$ = 88.5 N)

5.19 Suppose the rollers used in Q.18 above were changed giving coefficient of friction ranging from 0.20 - 0.65. Plot the graph of T$_{max}$ and T$_{min}$ for the values of coefficient of friction in 0.05 increments. Is there any difference in the slope of the two curves? For f = 0.27 what are the values of T$_{max}$ and T$_{min}$? What values of f are T$_{max}$ = 250 N and T$_{min}$ = 125 N? *(No difference; for f = 0.27, T$_{max}$ = 310 N, T$_{min}$ = 132.5 N; for T$_{max}$ = 250 N, f = 0.39; for T$_{min}$ = 125 N, f = 0.28)*

5.20 A belt conveyor is to be used to convey 80 t/h of egusi melon seeds at a speed of 2.5 m/s a distance of 12 m and to a height of 4 m for processing. If C$_o$ = 1.18, α = 180°, and f = 0.29, calculate the power required to drive the pulley shaft. If the pull factor is to vary from 2.5 to 5.0, how would the T$_{max}$ vary with it? What pull factor would give T$_{max}$ = 238 N? *(HP = 3.96 kW; pull factor = 3.1).*

5.21 If the loading speed of the rice paddy in Q.18 varies from 0.25 m/s to 2.0 m/s at 0.25 m/s increments, plot the curve of the total resistance force against the loading speeds with hopper friction coefficient = 1.35

5.22 At the loading station of a belt conveyor system, V$_o$ = 1.2 m/s; V = 2.5 m/s. If the load of 35 t/h of grain takes 3 sec to attain conveyor velocity, what is the K. E gained in the period, the distance travelled by the belt and the sliding distance of the load? Find also the work performed by the sliding load and the total resistance force at the loading station if ψ = hopper friction = 1.3 *(K.E = 23.4 kJ; d$_b$ = 5.55 m; d$_s$ = 1.95 m; Work = 8.21 kJ; F$_t$ = 16.43 N)*

5.23 In Q.22 above, if H = 13.5 m and L = 80 m, what will be the power and effective pull when C$_o$ = 1.3? *(HP = 11.23 kW; T$_e$ = 457.2 N).*

6

CHAIN CONVEYOR

6.1 INTRODUCTION

Chain conveyors are endless conveyors employing chains of various designs as both the driving traction element and, in some cases, carrying the whole weight of the materials to be conveyed. They are mechanical devices which may be fitted with special attachments (pendants, plastic chains) and used to carry, drag or scrape bulk materials from one location to another (Strobel et al., 2017).

Chain conveyors utilise a powered continuous chain arrangement, carrying a series of single pendants. The chain arrangement is driven by a motor, and the material suspended on the pendants are conveyed. Chain conveyors are primarily used to transport heavy unit loads, e.g. pallets, grid boxes, and industrial containers. These conveyors can be single or double chain strand in configuration. Many industry sectors use chain conveyor technology in their production lines (Kulkarni et al., 2018).

The type of **attachment** gives the conveyor its name; e.g. apron and pan, flight, drag, en-masse etc. The descriptions of some chain conveyors are given below.

a. **Drag Chain Conveyor** has an endless chain without attachment, which drags materials through a trough. Its operation relies on sliding friction between the conveyed materials and the smooth walls of the trough which causes the load to flow in compact unbroken stream. They have no crushing effect and are suitable to move frail materials, e.g. very fine, granular materials like wood chips, sand dust, pulpwood or log handling. Operating speed is between 0.1 to 0.6 m/s with low throughput capacities of 28 m³/h for 500 mm wide conveyor.

b. **Scraper or Flight Conveyor** is similar to the drag conveyor except that this has one or more strands of chain to which bars or flights made of malleable iron or plate are attached, which push materials through a trough. They are used for granular, non-abrasive materials and extensively for moving raw products, beets, potatoes, small grains into processing plants. This conveyor conveys materials batch-wise and may break the load in transit and therefore is used to handle unbreakable loads. However, with the flights submerged in the conveyed material and operating in an enclosed trough, the conveyed material moves as a continuous mass filling almost the whole cross-section of the trough. The inclination of flight conveyors should not be more than 30^0 to $45^{0,}$ otherwise, the capacity will be drastically reduced. Their operating speed is between 0.2 and 1.0 m/s and the trough width 3 to 4 times the flight height. Flights use roller chain while scrapers use sliding chain. They are widely used in agriculture, the chemical and mining industries. Their use is limited when handling moist, sticky materials. Breakages of load, intensive trough wear and high-power consumption also limit their use.

c. **En-masse conveyor** also termed the continuous-flow conveyor is adapted to convey fine, granular, crushed or pulverized, flaky, non-abrasive and non-corrosive free-flowing materials over a length of up to 100 m, a lift of up to 20 m and handling between 200 to 250 t/h at speeds varying between 0.1 to 0.6 m/s depending on the nature of load. The conveyor consists of a series of skeleton or solid U-shaped flights on an endless chain within a closely fitted casing, pipe or duct for the carrying run. The full cross section of the casing is used in conveying the materials horizontally, vertically or on an incline. Successful transport in an en-masse conveyor is dependent upon the frictional resistance between the material conveyed, the flight and the casing as well as the internal friction or shear strength of the material itself. The total enclosure of the en-masse conveyor ensures operational safety and meets environmental demands. It is suitable for combined processing applications such as cooling, heating, drying of conveyed products. Blended mixture of products can be moved without segregation (King, 1980).

d. **Tubular Drag Conveyor** is a continuous-flow flight conveyor with a series of discs mounted on an endless chain drawn through mild steel or stainless-steel tubes of circular cross-section ranging from 50 mm to about 250 mm in diameter. It covers a distance of some 30 to 50 m that may include bends to change its direction three or four times. Capacities may be up to 40 t/h with free-flowing granular materials moving with speeds of 0.03 to 0.6 m/s. Dry powdery chemicals, and dried industrial sludge and grit, common to sewage and water treatment processes can also be handled.

e. **Wide-Chain Drag Conveyor** is a modification of the en-masse conveyor whose moving element is a belt which may be perforated or not. It can handle dry free-flowing, fine and granular materials over distances of up to 50 m at a rate not more than 40 t/h along horizontal sections or on inclines not steeper than 12^0.

f. **Trolley Conveyor** consists of an overhead I-beam track with trolleys that are fastened together by chain. It is used for products of large unit size or those handled in boxes or baskets. Loads are usually hung from trolleys travelling overhead. Meat products, fruits and vegetables may be handled. Flexibility is an asset of this conveyor as it may be designed to make sharp turns up to 180^0 and up steep inclines. The conveyor is used for the performance of such processing operations as blanching, cooking, cooling, blast cleaning, impregnating, drying, etc. There are many modifications of the trolley conveyor, e.g. overhead push type and overhead tow conveyor. They carry loads of up to 2.5 t at speeds between 0.1 and 45 m/min. The loads may be up to 12 m long and the path between 500 and 600 m. Trolleys are spaced at a distance governed by the spacing of carriers and the necessity to provide for changes in elevation along the path.

g. **Apron Conveyor and Pan Conveyor.** These consist of chains to which close-fitting series of metal plates or pans or timber boards are attached to form a continuous moving bed capable of curving both vertically and horizontally. They can be used for conveying granular or lumpy or materials in sacks and materials of large unit size. They can handle heavy, abrasive and hot materials of over 200 t/h. Apron conveyors are available in widths from 200 mm to 2 m, and fitted with side flanges of up to 300 mm in height. Their operating speed varies between 0.05 and 0.6 m/s depending on application and load and can operate on inclines up to 25^0 only to prevent the conveyed materials from cascading. Even with cleats attached, the material's angle of repose must not be exceeded. Aprons carry rather than drag their load, hence less friction to overcome and therefore require less power than screw or scraper conveyors. Again, because they utilize roller chains, they require less pull or power. They are more efficient when used horizontally. They can be used as feeder conveyors and as an alternative to the belt conveyor used as a picking or grading table.

h. **Multi-flexing chain conveyor** systems use the plastic chains in many configurations. The flexible conveyor chain design permits horizontal as well as vertical change of direction.

6.2   COMPARING CHAIN AND BELT CONVEYORS

1. Belt conveyors are expensive, quiet, fast, and mechanically efficient and must be carefully engineered to ensure satisfactory performance.
2. Chain conveyors are not so expensive, may be noisy, slow, not mechanically efficient, and do not require much specialized skill for design.
3. Unlike belt conveyors, chain conveyors are adaptable to elevated temperatures and can handle coarsely broken materials.
4. Belts seem wider than chains but chain speed is much smaller than belt speed.
5. Chain conveyors compare favourably with belt conveyors when the load to be handled is heavy, lumpy or abrasive.
6. The power consumption of chain conveyor is considerably higher than that of belt conveyor, other things being equal.
7. Some chain conveyors have more difficulty in handling abrasive, moist and sticky materials as well as those with solid inclusions which may be jammed between flights and conduit walls than belt conveyors.
8. Chain conveyors can operate at larger inclines than belt conveyors. Steep inclines and vertical sections do not impair chain performance.
9. Apron conveyor is more costly per metre than a belt conveyor, but its use lends itself where a belt is not applicable.
10. Drag and scraper conveyors are higher in maintenance costs than other means of conveying.

6.3   CHAIN CONVEYOR DESIGN CONSIDERATIONS
6.3.1   Design Features

The driving traction element is the chain. Most installations are equipped with low-pitched rivet-less chains in which two inner plates of one link, interconnected by a bushing, are linked with the adjacent link consisting of two outer plates by means of a pin secured to the plates and passing through the bushing. The most stressed and wearing-out parts of a chain are the pins. The components of a chain are subjected to tensile and shock loads as well as fatigue and wear. The chain may be a sliding one or rolling one.

Chain sliding is simple in construction with few moving parts, effective in dirty applications and well suited for impact conditions. Chain rolling has smoother operation, less pulsation, lower friction than sliding chain which has a higher horsepower requirement. The rolling chain has smaller motors and lower operating cost. Chain-driven conveyors with roller surfaces are occasionally non-powered (gravity) or, more usually, powered (live) (Gunal et al., 1996). Such conveyors are relatively inexpensive, readily assembled and adjusted, and well suited for a wide range of loads, provided the materials being transported have, or are mounted on, a rigid riding surface (Allegri, 1992). These chain-driven conveyors may accumulate material by use of slip or clutch mechanisms built into the rollers. The "roller flight" variety typically uses two parallel sections of chain supporting rollers on non-rotating shafts. Those rollers can turn under the material, permitting its accumulation (Gould, 1993).

6.3.2   Tensile Loads
In designing a chain conveyor, the pull or tension required for moving the load or transmitting power is the basic force to be considered. Centrifugal force at the sprockets and catenary tension on the slack side all add to increase the total tension in the chain.

6.3.3   Wear
Wear in chain conveyor occurs: -
i   When pins oscillate in the bushing
ii   When bushings rub on the sprocket teeth
iii   On the outside diameter of rollers
iv   Between the rollers and the bushings
v   On the top bottom of the links
vi   When there is improper alignment

6.3.4   Shock Loads
Shock is a result of the loading conditions and the action between the chains and the sprocket. Shock loading increases as the pitch and the speed of the chain increase, and the sprocket size decrease. Shock loads must be brought to a minimum during design.

6.3.5   Fatigue
Fatigue becomes an important design factor for chain conveyors where the chains are operated beyond their rated capacity or a subject to high cyclic loading.

6.3.6   Oscillation of chain speed at sprockets
When the roller approaches the sprockets (Fig 6.1) it does not follow a tangent to the pitch circle, but moves in a series of arcs causing oscillation. This phenomenon sets up pulsations and alters the linear velocity of the chain.

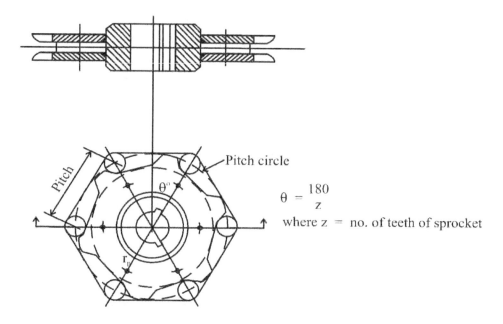

$$\theta = \frac{180}{z}$$

where z = no. of teeth of sprocket

Fig. 6.1: Driving chain wheels

The speed difference is given in Eqn. 6.1.

$$dV = \omega r_p\left(1 - Cos(180^o/z)\right) \qquad \qquad \dots 6.1$$

Where ω = radial speed of sprocket; r_p = pitch circle radius; z = number of teeth on sprocket
The greater the number of teeth, the less speed variation due to oscillation (see Fig.6.1). Excessive oscillation results in noisy operation.

6.3.7   Shape of Flights
Flight profiles must suit the type of material being conveyed. For horizontal conveying flat flights are used. They may be chamfers to counteract any tendency for the flight to fly out of the material. Skeleton flights are used on elevator applications.

6.3.8   Materials for Chains
Chains are made of quality carbon or alloy steel as well as malleable iron to ensure adequate strength and wear resistance. The chain provides a convenient means of attaching load-supporting members and running gear of the conveyor, positively transmits the drive imparted to it by the sprocket, and displays low stretchability under a load. However, failure of this chain is continuous problem which causes huge losses (ACA, 2006). Because of that, roller chain is the most important element (Kadam and Deshpande, 2015). So, the factor of safety used in selecting chains is based on the breaking strength.

6.3.9 Chain Conveyor Classes

Conveyors are classified according to type as could be seen in Table 6.1a. With the scraper type, materials slide on the tractive element with or without flights. While with the apron type, materials are carried on the chain whether they are sliding or rolling.

Table 6.1a   Classes of Chain Conveyor

Class	Chain	Material	Conveyor type
1	Sliding with or without flights	Sliding	Scraper, drag
2	Rolling	Sliding	Scraper
3	Sliding	Carried	Apron and pan
4	Rolling	Carried	Apron and pan

6.4   CALCULATIONS FOR CHAIN CONVEYOR DESIGN
6.4.1   Strength of Chain
Chains are selected by the relationships given as Eqns. 6.2 and 6.3.

$$S_b \geq kS_a \qquad \qquad \text{...6.2}$$

where S_b = breaking strength of chain (N); S_a = actual working load tension in the chain (N); k = safety factor; k = 6 - 7 for conveyors having no vertical or steeply inclined sections; k = 8 - 10 for escalators and steeply inclined conveyors.

Also

$$S_a = (1.6 - 1.8)S_c \qquad \text{(N)} \qquad \qquad \text{...6.3}$$

where S_c = calculated maximum chain tension.

6.4.2   Volumetric capacity
The volumetric capacity is principally a function of the cross-section area, A (m^2), of the bed of conveyed materials, the velocity of the chain, V (m/s), and a capacity factor as in the belt conveyor. The speed of the chain range between 0.03 m/s and 1.0 m/s depending on the nature of the material carried and given as Eqn. 6.4.

$$V_s = 3600AV\psi \qquad \text{(m}^3\text{/h)} \qquad \qquad \text{...6.4}$$

and the mass throughout capacity is as in Eqn. 6.5.

$$Q = 3600A\varrho V\psi \qquad \text{(t/h)} \qquad \qquad \text{...6.5}$$

where ϱ = density of material (t/m^3).

For Apron Conveyors

In the apron and pan conveyors (Fig.6.2a), the throughput capacity is given for an apron conveyor by Eqn. 6.6.

$$Q = 3600A\varrho V = 650B^2 V\varrho tan\theta_s \quad (t/h) \qquad\qquad ...6.6$$

for a pan conveyor (Fig. 6.2b) it is given as Eqn. 6.7.

$$Q = 3600Bh\varrho V\psi \quad (t/h) \qquad\qquad ...6.7$$

where: B = width of conveyor (m); h = height of pan (m); $\theta_s = surcharge = 0.4\theta_r$; $\varrho = density\ of\ material$; A = cross-sectional area (m²); V = speed of conveyor (m/s); $\psi = capacity\ factor$; b = 0.85B = width of conveyor with load; $\theta_r = repose\ angle$.

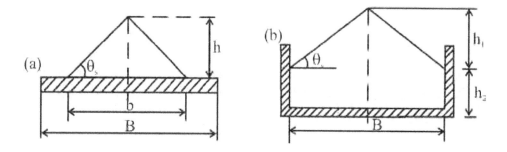

Fig. 6.2: Cross sectional loading of apron (a) and pan (b) conveyors

The capacity factor depends on the type of material loading as shown in Table 6.1b.

Table 6.1b   Capacity factor ψ for Apron conveyors

Mode loading	Capacity factor
High tonnage fed uniformly	0.85 - 0.90
High tonnage fed non-uniformly	0.75 - 0.80
Loads of variable size fed non-uniformly	0.50 - 0.70

For inclined apron and pan conveyors, Eqn. 6.6 and Eqn. 6.7 are multiplied by a flowability factor K (Table 6.2)

Table 6.2.   Flowability factor K for apron conveyors.

Inclined deg.	Apron Conveyor	Pan Conveyor
<10^0	0.93-1.0	0.97-1.0
10 - 20^0	0.86-0.92	0.93-0.96
>20°	<0.85	<0.93

The throughput capacity of a product of density 800 kg/m^3 moving at 0.05 m/s is given in Table 6.3

Table 6.3   Capacity of Apron conveyors.

Width between skirt plates (mm)	Capacity 800kg/m^3 at 0.05 m/s speed			
	Dept of load (mm)			
	305	406	508	610
610	559	762	-	-
762	660	940	1194	1422
914	864	1143	1422	1702
1097	992	1321	1651	2007

For flight Conveyor

The throughput capacity of flight conveyors is based on the batch-wise movement of the load in the conveyor with the flights submerged in the load. Fig. 6.3 shows the cross-sectional area of the material in front of the moving flight as a trapezoid, and the throughput capacity is given as Eqn. 6.8.

$$Q = 3600 Bh\psi V\varrho \quad \text{(t/h)} \quad\quad\quad \text{...6.8}$$

The cross-sectional area is given by Eqn. 6.9.

$$A = Bh\psi \quad \text{(m}^2\text{)} \quad\quad\quad \text{...6.9}$$

where Q = throughput capacity (t/h); B = width of the flight (m); h = height of the flight (m); $\psi = filling\ coefficient$; V = velocity of flight (m/s); $\varrho = density\ of\ conveyed\ material.$

The value of ψ ranges from 0.5 - 0.6 for granular materials and 0.7 - 0.8 for lumpy loads. ψ is the capacity factor expressing the ratio of the actual volume of conveyed material between a pair of flights to the geometrical one. The flowability factor K for inclined flight conveyors are given in Table 6.4

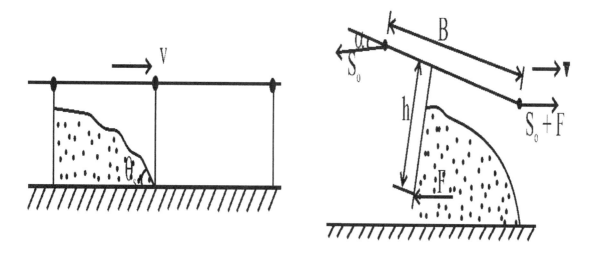

Fig. 6.3: Calculating flight conveyor capacity

Table 6.4   Flowability factor K for Flight Conveyors

Material	Incline, deg.					
	0	10	20	30	35	40
Granular	1	0.85	0.65	0.5	-	-
Lumpy	1	1	1	0.75	0.6	0.5

For En-masse conveyor (Fig.6.4)

The volumetric throughput capacity of an en-masse conveyor in m^3/h is calculated by Eqn. 6.10.

$$V_s = 3600AV\psi \hspace{3cm} ...6.10$$

ψ is the capacity factor which takes account of the cross-sectional area of the moving elements reducing the effective section of the conduit. Values of ψ for different transport situations are given below:

$$\psi = 0.83 - 0.99 \text{ for inclines } 0 - 20^0$$
$$\psi = 0.8 \hspace{1cm} \text{for inclines } >20^0 \text{ with lumpy loads}$$
$$\psi = 0.6 \hspace{1cm} \text{for granular loads}$$
$$\psi = 0.45 \hspace{1cm} \text{for fine materials}$$

Optimum velocities of the chain and flights depend on the nature of the conveyed material. Free-flowing particulate and granular materials have $V > 0.5 \, m/s$; abrasive materials and aerated (cement) materials have $V < 0.25 \, m/s$; fibrous and flaky products have $V \leq 0.4 \, m/s$.

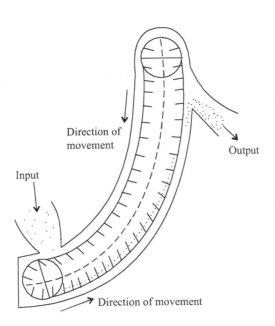

Fig. 6.4: En-masse conveyor

6.4.3 Estimating the total required chain pull

The following forces are to be overcome by a chain conveyor moving materials: -

i.   The rolling resistance of moving conveyor parts in a horizontal section
ii.   The friction in sprocket bearings
iii.   The resistance force due to chain stiffness
iv.   The impact loads in chain conveyor
v.   The tensions T_1 and T_2 in tight and slack sides of the chain
vi.   The force required to move the chain and load up an incline

The estimation of some of these forces are given below:

i. The aggregate load (F_1) sustained by conveyor rollers horizontally is Eqn. 6.11.

$$\sum k = (q + q_o)gL \quad \text{(N)} \qquad \qquad \ldots 6.11$$

where q = weight of conveyed material per unit length (kg/m); q_o = weight of chain and attachment per unit length (kg/m); L = horizontal centre distance (m).

The rolling resistance is given as Eqn. 6.12.

$$F_1 = C\sum k = CLg(q + q_o) \quad (N) \qquad \ldots 6.12$$

where C = friction factor for conveyor rollers (Table 6.5).

Table 6.5   Friction Factor C for Rivet-less Roller Chains

Grade of Service	Friction Factor Depending on type of Roller Bearing	
	Plain	Rolling contact
Light Moderate Heavy	0.06 - 0.08 0.08 - 0.10 0.10 - 0.13	0.020 0.030 0.045

For trolley conveyors: -
      C = 0.02 - 0.04 for straight sections
      C = 0.04 - 0.08 for traction-wheel turns on plain bearing
      C = 0.02 - 0.06 for traction-wheel turns on rolling contact ones
      C = 0.02 - 0.05 for roller turns and vertical curves.

For apron conveyors friction of material against skirt-boards is added to Eqn. 6.12 to give Eqn. 6.13.

$$F_{load} = f h^2 \varrho K_i L \qquad (t) \qquad \ldots 6.13$$

where $f = coefficient\ of\ sliding\ friction$ (Table 6.6); h = effective height of skirt-board (m) $\varrho = bulk\ density\ of\ materia$ (t/m^3); V = chain speed (m/s); $\theta_r = angle\ of\ repose$; L = length (centre-to-centre) in m; $K_i = /(V^2 + 1.2)/(1 + Sin\theta_r) = coefficient\ due\ to\ internal\ friction\ on\ skirt - board$

For flight conveyors, there is friction between the side walls and bottom and higher friction factors are required as shown in Eqn. 6.14.

$$C_{load} = 1.1f \quad (approx.). \qquad \ldots 6.14$$

where f = coefficient of material against trough; C = 0.10 - 0.13 for plain bearing rollers; C = 0.25 - 0.40 for sliding-type chains

The aggregate resisting force for flight conveyors is given as Eqn. 6.15.

$$F = (qC_{load} + q_oC)gL \qquad \text{(N)} \qquad\qquad ...6.15$$

where C = friction factor for the flights; C_{load} = friction factor allowing for internal friction and friction between the material and trough; L = length of trough.

Table 6.6   Friction Coefficient (sliding)

Chain sliding on steel track (unlubricated)	0.3-0.5
Chain sliding on steel track (lubricated)	0.2
Chain sliding on steel track Hard Wood	0.5-0.6
Chain sliding on steel track Plastic wear strips	0.2-0.3
Chain sliding on steel track Ultra high molecular polyethylene	0.15-0.2
Oak on Oak (parallel fibres)	0.48
Oak on Oak (cross fibres)	0.32
Cast iron on mild steel	0.23
Mild steel on mild steel	0.57
Grain on rough board	0.30-0.45
Grain on smooth board	0.30-0.35
Grain on iron board	0.35-0.40
Woodchips, pulp logs on steel	0.4
Gravel (dry) on steel	0.45
Gravel (run of bank)	0.6

(ii)   The resisting force (F_2) due to the sprocket bearing friction is given as Eqns. 6.16 and 6.17.

$$F_2 = F_b f\, d_o/D_s \qquad \text{(N)} \qquad\qquad ...6.16$$

where F_b = resulting force on sprocket shaft bearing calculated from T_1, T_2 and weight of sprocket-and-shaft assembly (N); f = coefficient of friction for journals; d_o = diameter of shaft journal (m); D_s = diameter of rolling circle of sprocket (m).

Usually, $F_2 = (0.03 - 0.05)T_1 \qquad \text{(N)} \qquad\qquad ...6.17$

(iii)   The resisting force (F_3) due to chain stiffness or articulation is given by Eqn. 6.18.

$$F_3 = (T_1 + T_2)f_o \, d/D_s \qquad (N) \qquad \qquad \qquad ...6.18$$

where $f = 0.5 - 0.6$ = coefficient friction for pivot and contacting surfaces of rollers and sprocket teeth; d = pivot diameter (m).

Depending on the wrap angle, the sum $F_2 + F_3 = F_{s.b}$ is given by Eqn. 6.19.

$$F_{s.b} = 0.05T_1 \text{ for } 90^0 \text{ wrap angle}$$
$$F_{s.b} = 0.07T_1 \text{ for } 180^\circ \text{ wrap angle} \qquad \qquad ...6.19$$

where $F_{s.b}$ = total friction in bearing.

(iv)   The pulsating motions of the chain links when going around the sprockets give rise to impact loads which increase with chain pitch and speed higher than 0.2 m/s. Thus, there is an increase in chain pull and chain fatigue.

      For drive sprockets operating at constant angular velocity, the maximum chain acceleration is given by Eqn. 6.20 and Eqn. 6.21.

$$j_{max} = \pm \frac{\omega^2 t}{2} \qquad \qquad \qquad ...6.20$$

Since $\omega = \pi n/30$ *and* $n = 60 V_{ch}/Zt$

then $j_{max} = \frac{2}{t}[\pi V_{ch}/Z]^2 \qquad (m/s^2) \qquad \qquad ...6.21$

where ω = constant angular velocity of drive sprocket; V_{ch} = circumferential speed of chain (m/s); t = chain pitch (m); Z = number of sprocket teeth.

As soon as a sprocket engages the next articulation of the chain, the acceleration instantaneously increases two-fold. Therefore, the impact load becomes $2mj_{max}$ in which m the modified weight of moving components and load is given by Eqn. 6.22.

$$m = (q + \psi q_o)L \qquad (kg) \qquad \qquad ...6.22$$

where ψ = factor for accelerating less than j_{max}, stiffness and sag; L = conveyor length (m); $\psi = 2 \text{ when } L < 25m$; $\psi = 1.5 \text{ when } 60 \leq L < 25$; $\psi = 1 \text{ when } L > 60m$

The chain pull resulting from the instantaneous impact load equals that set up by the two-fold static load ($4mj_{max}$) minus the force of inertia (mj_{max}) given as Eqn. 6.23 and Eqn. 6.24.

$$P_{impact} = 4mj_{max} - mj_{max} = 3mj_{max} \qquad\qquad \ldots 6.23$$

$$P_{impact} = 6L[((q + \psi q_o)/t)((\pi V_{ch})/Z)^2] \qquad\qquad \ldots 6.24$$

6.4.4 Estimation of Total Required Chain Pull for Different Classes of Chain Conveyor

In estimating the total required pull of the chain conveyor, consideration is given to both class of the conveyor and the way it is being loaded as well as its inclination to the horizontal (Fig. 6.5). The way the load is being conveyed (whether sliding or carried) is another important factor to be considered.

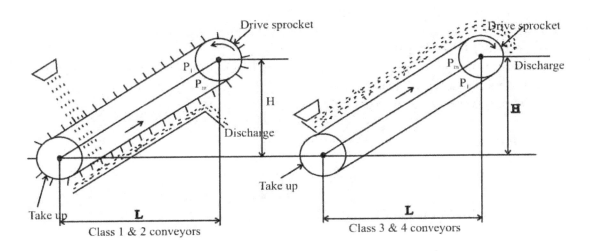

Fig. 6.5: Methods of conveyance by classes of chain conveyor

A. For Class 1 and 2 conveyors (Table 6.1a), the chain is sliding or rolling and the materials are sliding. When they are horizontal (i.e. $H/L < f_1$) we have Eqn. 6.25.

$$P_m = L(2f_1 q_o + f_2 q + f_2 \varrho h^2) + qH \qquad \text{(kg)} \qquad\qquad \ldots 6.25$$

When they are inclined (i.e. $H/L > f_1$), we have Eqn. 6.26.

$$P_m = L(f_1 q_o + f_2 q + f_2 \varrho h^2) + H(q_o + q) \quad \text{(kg)} \qquad\qquad \ldots 6.26$$

B. For Class 3 and 4 conveyors (Table 6.1a), the chain is sliding and rolling and the materials are being carried. When they are horizontal (i.e. $H/L < f_1$) we have Eqn. 6.27.

95

$$P_m = f_1L(2q_o + q) + qH + \varrho f_2h^2L \qquad \ldots 6.27$$

When they are inclined (i.e. $H/L > f_1$) we have Eqn. 6.28.

$$P_m = (q_o + q)(f_1L + H) + \varrho f_2h^2L \qquad \ldots 6.28$$

where the take-up force $\quad = \quad 2f_1q_oL \qquad \ldots 6.29$

f_1 = coefficient of friction of chain on runway (0.15 - 0.5); f_2 = coefficient of friction of material on trough (0.15 - 1.0); ϱ = material bulk density (t/m³); q_o = wt. of moving parts - chains, flights, slats etc. (kg/m) = (0.10 - 0.15% of total conveyed load); h = height of material rubbing against the trough (m); P_m = total static chain pull (kg); T_1= slack side tension (kg); V = conveyor speed (m/s); Q = throughput capacity (t/h); L = horizontal distance (m); H =vertical rise (m); q = wt. of material handled per metre of conveyor (kg/m) = $Q/3.6V$ (kg/m) .

6.4.5 POWER DETERMINATION

Calculation of the power requirement involves estimation of the various frictional losses in the chains, rollers, guide-ways, driving mechanisms and power unit. With the chain tension estimated, the horsepower is given by Eqn. 6.30.

$$HP = P_m V/\eta \quad horizontal \ conveyor;$$

$$HP = ((P_m \pm T_1)/\eta)V \quad inclined \ conveyor \qquad \ldots 6.30$$

where η =mechanical drive efficiency (0.80 – 0.87); V = chain speed (m/s).

The above calculations for pull (P_m) are for static and not dynamic conditions. For dynamic conditions we consider load fluctuations and daily operating period of which we provide the following service factors (Table 6.7) as well as chain speed. Though the speed is depending on the conveyed materials, chain speed is kept low to minimize the wear and tear on the conveyor (Table 6.8). A combination of the chain speed and the number of teeth on the sprockets is covered by the speed factor (Table. 6.9). Higher speed factors are used for chains made with cast than those made with steel materials.

Therefore, to calculate the design working load for the chain conveyor we use Eqn. 6.31 and Eqn.6.32.

For single strand conveyor, the design working load is given as Eqn. 6.31.

$$Design \ working \ load \ = \ P_m \ x \ service \ factor \ x \ speed \ factor \qquad \ldots 6.31$$

For a multiple strand conveyor, the design working load is given as Eqn. 6.32

Design working load =

P_m x service factor x speed factor x 1.2/No. of strands   ...6.32

Table 6.7:   Service Factors

Type of load	Operating Conditions		Daily Operation Period	
	Start, stop frequency under load	% load added at a time	8-10hrs	24hrs
Uniform	Less than 5/day	Less than 5%	1.0	1.2
Moderate peaks	5/day - 2/hrs	5 - 20%	1.2	1.4
High peaks	2/hr - 10/hr	20 - 40%	1.5	1.8

Table 6.8:   Some Typical Speeds

Materials	Speed
Very abrasive: coke, ore, bauxite, silica	0.025 - 0.031 m/sec.
Moderately abrasive: coal, limestone, salt, etc.	0.31 - 0.36 m/sec
Mildly abrasive: grains, corn, soybeans, woodchips, etc.	0.51 - 1.02 m/sec.

The effective tension in the chain, is given in Eqn. 6.33

$$T_{max} = 1.25 \; x \; Design \, working \, load.$$
$$T_{min} = 0.25 \; x \; Design \, working \, load. \qquad ...6.33$$
$$T_e = T_{max} - T_{min}$$

The new design power of the chain will be calculated using the effective tension in the chain (Eqn. 6.34).

$$HP = T_e V/\eta \qquad (kW) \qquad\qquad ...6.34$$

Table 6.9:   Speed Factor

No of teeth on sprocket	Chain Speed (m/s)							
	0.25		0.51		0.76		1.02	
	Cast	Steel	Cast	Steel	Cast	Steel	Cast	Steel
6	1.6	1.4	2.3	2.0	3.3	2.9	5.0	4.4
7	1.3	1.1	1.6	1.4	2.0	1.8	2.6	2.3
8	1.2	1.0	1.4	1.3	1.7	1.5	2.0	1.8
9	1.1	1.0	1.3	1.2	1.6	1.4	1.8	1.6
10	1.0	0.9	1.3	1.1	1.4	1.2	1.6	1.4
11	1.0	0.9	1.2	1.0	1.3	1.2	1.5	1.3
12	1.0	0.9	1.1	1.0	1.3	1.1	1.4	1.2
14	1.0	0.8	1.1	0.9	1.2	1.0	1.3	1.1
16	0.9	0.8	1.0	0.9	1.1	1.0	1.2	1.0
18	0.9	0.8	1.0	0.9	1.0	0.9	1.2	1.0
20	0.9	0.8	1.0	0.9	1.0	0.9	1.1	1.0
24	0.9	0.8	0.9	0.8	1.0	0.9	1.1	0.9

6.4.6   Cautions in Flight - Conveyor Selection

In selecting this type of conveyor, it is necessary to consider the following cautions for the proper operation of the conveyor:

1. With abrasive materials, the trough is designed to provide a replacement of bottom plate without disturbing side plates
2. For inclined conveyors which can slide back when stopped under the load, a solenoid brake or other automatic backstop device is provided.
3. For steeply inclined conveyors, flights should be closely spaced to avoid the cascading of materials over the top of the flights.

6.5   OPERATION AND MAINTENANCE OF CHAINS AND SPROCKETS

When in operation, a hard steel chain continuously slides over a hard steel bottom, causing both to wear. The wear on both the sprocket and the chain depends on the following:

  i. Kind of product carried
  ii. Kind of construction material for chain/sprocket
  iii. Speed of the chain
  iv. Idle time of the conveyor

So, it is important that for preventive maintenance of the chain conveyor the following details must be checked:

i. Chain tension, chain stretch and tensioning device
ii. Links, flights, buckets and connecting pins
iii. Elements of the return part of the chain (rolls, sliding strips or plate)
iv. Drive and return sprockets and their engagement of the chain links.
v. Greasing of all bearings,
vi. Alignment of the trough.
vii. Complete drive mechanism (motor, gearbox, couplings)
viii. Motion control and the overflow valve with its limit switch

A more detailed maintenance will have to deal with wear, tension and environmental factors as enumerated below:

6.5.1 Wear

A badly worn sprocket will readily damage the chain rapidly and excessively, and vice versa. So, it is being suggested that we do the following:

i. Inspect sprocket frequently and improve the life, by lubrication, if excessive wear is taking place use correct sprocket for the chain in terms of type, size, and material of construction.
ii. Keep chains as clean as possible all the time.
iii. Replace sprockets when installing a new chain.
iv. Replace damaged links as soon as observed.
v. Align trough sections properly to reduce excessive wear, noisy operation and damaged flights.

6.5.2 Tension

The proper tensioning of the chain is important (Glugosh and Knotek, 2019), use the tensioning device to give enough tension to keep the chain in contact with the sprocket. Too much tension will overload the system as too little tension will cause jerks, slaps and slips.

6.5.3 Environmental Conditions

Chains and sprockets work best in good environments.

i. In corrosive conditions, vital parts are plated or made of more corrosion resisting material.
ii. Thorough lubrication (oiling or greasing) should be done regularly.
iii. Chains and sprockets should be operated within recommended load and speed limits, the

higher the speed the more the wear.

iv. Liner materials should be such that they wear faster than the chain.

v. In an abrasive environment, increase the hardness of the chain or liner and lubricate even with water. This is because wear to metal parts increase almost linearly with the abrasiveness of the material.

vi. Don't allow the conveyor to run empty, it causes increased wear on the trough bottom

vii. Check tightness of the loading and discharge slide gates as well as covers to avoid emission of dust.

6.6 SAMPLE PROBLEMS

Q.6.1 Design a steel flight conveyor to move 9.4 t/h of grain up a 30^0 inclined to a height of 6 m with a speed of 0.51 m/s. Assume density of grain 0.8 t/m^3, height of flight 12 cm. No. of sprockets = 12. Assume $\psi = 0.55$. (Class 1 conveyor).

Data: $Q = 9.4\,t/h; \varrho = 0.8\,t/m^3; H = 6\,m; \theta = 30^0; V = 0.51\,m/s; h = 0.12\,m; \psi = 0.55, f_1 = 0.25, f_2 = 0.4, q_o = 2.1\,kg/m, q = 0.76\,kg/m$

Solution: For flight conveyor, area $A = Bh\psi = B \times 0.12 \times 0.55 = 0.066\,B$

Volumetric capacity $= Q/\varrho = 9.4/3600 \times 0.8 = 3.264^{-3}m^3 = AV = Bx0.12x0.55x0.51$

Flight width B $= 0.09697\,m = 0.10\,m$. Use **B = 10 cm**

Theoretical length of run of conveyor L $= 6/\sin30^0 = $ **12 m**

Additional length must be used to provide clearance and overhang at the top or discharge end. Let's add 2 m

∴ Total length $=$ **14 cm**

Total actual height $=$ **7 m**

Since the flight conveyor is a class 1 conveyor, then using Eqn. 6.26, the static chain pull is given by: -

$$P_m = L(f_1q_o + f_2q + f_2\varrho h^2) + H(q_o + q) \qquad \text{... 6.26}$$

Substituting the following values; $f_1 = 0.25$; $f_2 = 0.40$; $\varrho = 0.8$ t/m^3; h = 0.12 m; q = 0.76kg/m; q_o= 2.1 kg/m; H = 7 m; L = 7/tan30^0 = **12.12 m**

$P_m = 12.12\,(0.25 \times 2.1 + 0.40 \times 0.76 + 0.40 \times 0.8 \times 0.12^2) + 7(2.1 + 0.76)$
$= 12.12 \times 0.834 + 7(2.86) = 30.13\,kg = $ **295. 56 N**

Because of the flight width, two chains will be required and operating tension (pull) in each = 147.78 N.

For dynamic conditions, service factor (Table 6.7) = 1.2, Speed factor (Table 6.9) =1.0 and Mechanical drive efficiency η = 0.85.No. of strands = 2. Using Eqn. 6.32,

Design working load = 295.56 x1.2 x1.0 x1.2/2 = **212.80 N**

Using Eqn. 6.33,

T_{max} = 1.25 x design working load = 1.25 x 212.80 = **266 N** = T_i

T_{min} = 0.25 x design working load = 0.25 x 212.80 = **53.2 N** = T_2
T_e = T_{max} − T_{min} = **212.8 N** = Design working load

The horse power is given by Eqn. 6.34.

$$HP = T_e V / \eta = 212.8 \times 0.51/0.85 = \mathbf{127.68\ W}$$

A ½ HP motor would be suitable.

Q.6.2 A horizontal apron conveyor of width 165 mm conveys grains of density 1.34 kg/m³ at a speed of 0.48 m/s. If the capacity factor is 0.90, find the throughput capacity of the conveyor. What will be the horsepower required to drive the conveyor if the static chain pull is 1320 N and the motor efficiency = 0.85. Assume h = 50 mm and $\theta_s = 30^o$.

Data $B = 0.165\ m;\ \varrho = 1.24\ kg/m^3;\ V = 0.48\ m/s;\ \psi = 0.90;\ P_m = 1320\ N;$
$\eta = 0.85;\ h = 0.05\ m, \theta_s = 30^o = 0.4\theta_r.$

Solution: Using Eq. 6.6, $Q = 3600A\varrho V = 650B^2 V \varrho tan\theta_s$

$Q = 650 \times 0.165^2 \times 0.48 \times 1.24 \times Tan\ 30^o = 6.08\ kg/s = \mathbf{21.89\ t/h}$

Using Eq. 6.30, $HP = P_m V/\eta$ *horizontal conveyor* =1320 x 0.48 / 0.85 = **745.4 W**

Q.6.3 For a flight conveyor, $q = 65\ kg/m; q_o = 8\ kg/m; L = 15m; C = f_1 = 0.18; C_{load} = f_2 = 0.22$, find the aggregate resisting force. What will be the total required chain pull and the take-up force if $h = 0.09\ m, H = 5\ m; \varrho = 0.88t/m^3$.

Data: $q = 65\ kg/m;\ q_o = 8\ kg/m;\ L = 15\ m;\ C = 0.18, C_{load} = 0.22;\ h = 0.09\ m;\ H = 5\ m;\ \varrho = 0.88t/m^3.;\ L_{inc} = \sqrt{(15^2 + 5^2)} = 15.81m$

Solution: Using Eqn. 6.15 for inclined conveyor, $F = (qC_{load} + q_o C)gL_{inc}$

$F = (8 \times 0.18 + 65 \times 0.22) \times 15.81 = \mathbf{248.85\ kg.}$

Using Eqn. 6.28,

$$P_m = (q_o + q)(f_1 L + H) + \varrho f_2 h^2 L$$

$$= (65 + 8)(15 \times 0.18 + 5) + 0.88 \times 1000 \times 0.22 \times 0.09^2 \times 15 = \mathbf{585.62\ kg.}$$

Using Eqn. 6.29, Take-up force = $F_1 = 2f_1 q_o L = 2 \times 0.18 \times 8 \times 15 = \mathbf{43.2\ kg}$

Q.6.4 If the conveyor in Q.6.3 has 3 strands and service and speed factors of 1.4 each, find the effective tension in the chain and the design horsepower. (*Take* $\eta = 0.80$ *and* $V = 0.36\ m/s$)

Data: $P_m = 585.62\ kg$; *service factor* = *speed factor* = 1.4; $\eta = 0.80$; $V = 0.36\ m/s$; *no. of strands* = 3.

Solution: Using Eqn. 6.32,

Design working load = $P_m \times$ *service factor* $\times$ *speed factor* $\times 1.2/No. of strands*

$$= 585.62 \times 1.4 \times 1.4 \times 1.2 / 3 = \mathbf{459.13\ kg}$$

Using Eqn. 6.33, $T_{max} = 1.25 \times Design\ load = 1.25 \times 459.13 = \mathbf{573.91\ kg}$

$$T_{min} = 0.25 \times Design\ load = 0.25 \times 459.13 = \mathbf{114.78\ kg}$$

$$T_e = T_{max} - T_{min} = effective\ tension = \mathbf{459.13\ kg}$$

Using Eqn. 6.34, $HP = T_e V / \eta = 459.13 \times 9.81 \times 0.36 / (0.8 \times 1000)$
$$= \mathbf{2.027\ kW}$$

CHAPTER 6 Review Questions

6.1   What is a chain conveyor and enumerate the areas of use in agricultural engineering processing?

6.2   Describe the following types of chain conveyors: - drag (tubular, wide-chain), scraper or flight, en-masse, trolley, apron and pan.

6.3   Make a brief comparison between a chain and a belt conveyor.

6.4   Enumerate the parameters to be considered in chain conveyor design.

6.5   Enumerate the resistance to be overcome by a chain conveyor moving material and give the general equations of the total static chain pull in a horizontal and an inclined situation?

6.6   What factors do you include when calculating the design working load for the chain conveyor and why?

6.7   Enumerate the factors to be considered when selecting a flight conveyor for proper operation.

6.8   For the preventive maintenance of the chain conveyor, what must you do?

6.9   Enumerate the things you will do when you want to do a detailed maintenance of a chain conveyor.

6.10   A chain has a calculated max tension = 150 N, it moved round a 12-tooth sprocket with a radial speed of 0.6 m/s. If the pitch circle radius is 400 mm, find the speed difference of the chain and the breaking strength of the chain. Assume $K = 6.6$ and $K_1 = 1.6$. *(dV = 8.18 x 10^{-3} m/s; S_b = 1.6 kN)*

6.11   A horizontal apron conveyor has B = 800 mm, θ_r = 40°, V = 0.5 m/s. If the load of density 1.2t/m³ is being transported, calculate its throughput capacity. Assume ψ = 0.85 and K = 1.0. What is the load/width? *(71.57 t/h; 0.68 kg/m)*

6.12   The apron chain conveyor has q = 4.5 kg/m, q_o = 0.65 kg/m; L = 25 m, C = 0.085 and f = 0.35, find the aggregate friction of the material against the skirt boards. *(h = 0.097 m; K = 1.61; F = 107.92 N).*

6.13   If the apron conveyor in Q. 11 above has d_o = 10 mm; D_s = 720 mm, T_1 = 220 N, T_2 = 640N for 180° wrap angle, what is the total friction in bearing? *(F_2 = 8.8 N; F_3 = 15.4 N).*

6.14   If the above apron conveyor has V_{ch} = 0.5 m/s, t = 320 mm, Z = 12, find the total chain pull resulting from instantaneous impact load. *(M = 192.5 kg; J_{max} = 0.107 m/s²; P_{impact} = 20.6 N)*

6.15   If the conveyor in Q. 12 is of Type 3 and is inclined with H = 10m, what is the total static force and the horsepower required if now T_1 = 250 N, V = 0.45 m/s, ϱ = 0.80 kg/m³ and η = 0.85? Assume f = f_1 = f_2. *(P_m = 938.7 N; HP = 364.6 W)*

7

███████████████████████████

BUCKET CONVEYORS/ELEVATORS

7.1   INTRODUCTION

These are installations for carrying bulk materials in a vertical or inclined path in grain handling processing systems. Bucket elevator has evolved as advanced material handling equipment in the mechanized bulk material handling industry. The effective use of different types of bucket elevators completely depends on its design and types of bulk material (Mg Than Zaw Oo et al., 2019).

They consist of endless belt, chain or chains (traction elements) to which are attached elevator buckets; a single or double casing which serves to enclose or partially enclose the moving buckets; the head machinery at the upper end (belt pulley or chain wheel or sprocket to turn the traction element and discharge chute); and the boot machinery at the lower end (belt pulley or chain wheel or sprocket, a tensioning device, and means of feeding the material to be conveyed). Bucket conveyors are similar to apron conveyors (Chapter 6) except that bucket-shaped receptacles take the place of flat apron plates. As the buckets prevent the material from sliding back, the carriers can operate on inclines of up to 70^0.

Vertical bucket elevators are preferred as they eliminate the use of awkward casing arrangement and angle tracks for the return run. The elevator casing is made in sections and must be dust-tight, 4.8mm sheet metal or aluminium should be used. However, it must have a large cross-section and chain-guide to prevent sway contact between the buckets and the casing.

The belt-bucket elevator has the following main advantages over the chain-bucket type: -

i.   Higher speeds and higher capacities
ii. Smoother and quieter operation
iii. High abrasive resistance
iv. High resistance to corrosion

Bucket conveyors, in general, offer the following advantages: -

i.   Transportation on a single machine (without reloading) along horizontal or vertical paths
ii. Ease of intermediate unloading along the entire length of horizontal sections
iii. Possibility of conveying hot goods
iv. Good preservation of conveyed materials
v.   Possibility of conveying separately different kinds of loads

In general, bucket elevators are used to elevate any bulk material that will not adhere to the bucket. Belt-bucket elevators, operating best at up to 30^0 inclines to the vertical, handle most efficiently abrasive materials, which would produce excessive wear on chains. Chain-bucket elevators are used with perforated buckets when handling wet materials, to drain off surplus water. Belt-bucket elevators are advantageous for grain, cereal etc., if the temperature is not high enough to scorch the belt (i. e. below 121^0 C).

7.2   TYPES OF BUCKETS

The types of buckets used in bucket conveyors depend on the application. However, the tendency of materials to pack in the elevator boot decides bucket choice (Fig.7.1). Buckets are shaped with either sharp or rounded bottoms to facilitate discharge.

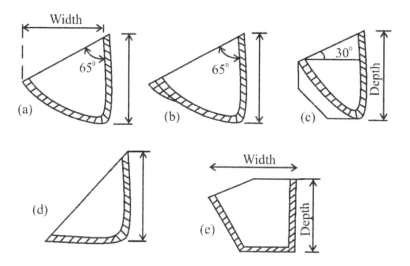

Fig. 7.1: Various types of buckets

The proper choice of a bucket informs the selection of the bucket elevator itself.

i.   Free-flowing non-packing materials, e.g. grains, are handled by high-front round-bottom buckets (Fig.7.1a).
ii.   With extra protection of the digging lip of the bucket, it could now be used for abrasive materials (Fig.7.1b).
iii.   Continuous-bucket elevators employ flat-front or V-shaped buckets with side guides (Fig.7.1c).
iv.   Sticky loads which tend to pack in the bottom of the bucket are conveyed using open buckets of the flat-bottom configuration (Fig.7.1d).
v.   High-front lips back-hooded buckets (Fig.7.1e) are used to maximize volumetric capacities and manipulate the material at the discharge point successfully over the preceding bucket.

vi.   With air escape holes (i.e. vented buckets) to control entrapped air, the grain industry has successfully handled flour and like materials.

vii.   For large-lump materials, extra-larges buckets are usually used and operated at an incline to improve feeding and discharge conditions.

7.3   METHODS OF LOADING THE BUCKETS

Loading is either by scooping the material on the bottom of the elevator (Fig.7.2a) or by pouring the material into the bucket (Fig.7.2b). In practice, bucket filling is likely by a combination of both methods (Fig. 7.2c). Scooping is used for conveying fine-sized lumps of granular, and/or dry easy flowing, pulverized materials, which offer little resistance to the buckets, e.g. milled peat, grain, cement, soil, sawdust etc. They can be scooped at speeds of 0.8 to 4.0 m/s using widely spaced buckets. Materials having high moisture contents and thus free-flowing can be scooped.

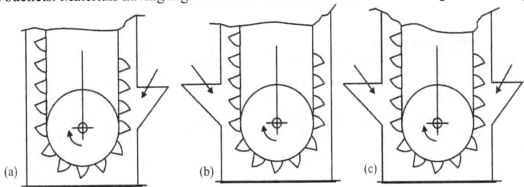

Fig. 7.2: Methods of loading bucket conveyors

Large-sized lumpy, coarsely broken and abrasive loads (gravel, ore, coal, etc.) are difficult to scoop and are therefore poured directly through loading chutes into upward moving closely spaced buckets moving with reduced speeds of not more than 1.3 m/s to minimize the bouncing or splashing of the produce from the bucket.

7.4   METHODS OF UNLOADING AND PRINCIPAL TYPES OF BUCKET ELEVATORS

There are three methods of unloading bucket elevators/conveyors in agricultural engineering practice. The different types of bucket conveyors except for the overlapping buckets and profiled-belt elevators are differentiated by these unloading or discharge methods. The characteristics of this discharge are important to the proper design and operation of bucket elevators.

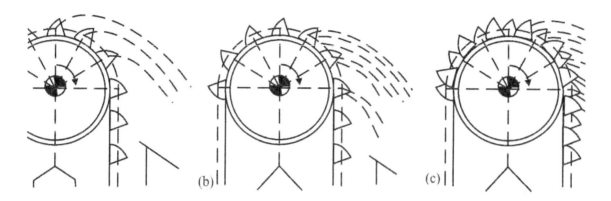

Fig. 7.3: Types of bucket elevator discharge

7.4.1 Centrifugal discharge elevators

These are the most common types for vertical discharge of free-flowing materials like grains. They are high-speed (1.3 to 3.5 m/s) elevators with uniformly but widely spaced buckets placed to prevent interference of the material during discharge by centrifugal force at the head of the elevator when the buckets revolve over the pulley or sprocket. It is very important to choose correctly the rotational speed of the pulley or sprocket and the position of the receiving spout relative to it (Fig.7.3a). Materials are fed by scooping action of the bucket at the boot or by direct loading into the bucket (Millier, 1958).

7.4.2 Positive (free gravity) discharge elevators

For light, dusty, sticky, fluffy or wet materials (feeds, cotton, salt, etc.) which tend to pack in the bucket and difficult to discharge, a modified arrangement (known as positive discharge) is used. It has buckets suspended at intervals between two chains. It has a knuckle or snub shaft (Fig.7.3b) under the head shaft sprocket or pulley which inverts the buckets with slight jolting so that discharge into the spout is by gravity. This method is less speedy (0.6 - 0.8 m/s) and employs larger but more closely spaced buckets than centrifugal discharge elevators. When snubs are not used, buckets of the flat-bottom type are used and the front wall of the preceding bucket serves as the guide for the material discharged from the succeeding one. Materials are fed into the buckets by scooping.

7.4.3 Continuous (directed-gravity) discharge elevators

These elevators had their buckets mounted continuously (not spaced) on a chain or belt (Fig.7.3c) and used primarily for handling large, heavy, lumpy and gritty materials too difficult to handle with centrifugal discharge machines. Materials are fed directly into the buckets through a chute situated 3 or 4 buckets above the boot. Discharge is largely by gravity. As the bucket revolves on the head pulley or sprocket, the material pours from the bucket onto the back surface of the preceding bucket and is directed by this and by the side walls of the bucket into the unloading spout of the elevator. Inclined ones allow the discharge chute to be placed directly below the discharging buckets for cleaner operation. The bucket speed is low (0.4 - 0.8 m/s) and suitable also for friable products and for those that are very fine, light and fluffy.

7.5   BUCKET CONVEYOR/ELEVATOR DESIGN CONSIDERATIONS

Optimum performance of bucket elevators is obtained by paying attention to its design in relation to the characteristics of the product to be conveyed, the required throughput, height of vertical lift and space available below feed point and above discharge point. The principal variables in the design being:

i	The bucket size and pitch
ii	The belt or chain speed
iii.	The diameter of the head and tail pulleys or sprockets.

7.5.1 Shape of Bucket

The design feature most strongly influenced by load characteristics and required throughput capacity is the type of elevator and shape of the bucket to be selected.

The weight of a bucket full load is given by Eqn. 7.1.

$$G = i_o \varrho \psi \qquad \qquad ...7.1$$

and the throughput capacity by weight, Q t/h of a bucket elevator travelling at V m/s with buckets spaced t_b metres apart is given by Eqn. 7.2.

$$Q = 3.6G \, V/t_b \qquad \qquad ...7.2$$

Substituting Eqn. 7.1 in Eqn. 7.2 and transposing, results in Eqn. 7.3.

$$i_o/t_b = Q/3.6V\varrho\psi \qquad \qquad ...7.3$$

where i_o = struck or geometrical volume of bucket (litre); t_b = pitch of buckets (m); V = speed of belt or chain (m/s) (Table 7.1); ϱ = density of material (t/m^3 or kg/1); ψ = coefficient of bucket filling. (Table 7.1).

For widely spaced buckets ψ= (2 to 3) D.
For closely spaced buckets (flats front with guiding side walls), ψ = D; where D = depth of bucket (m).

Table 7.1: Speed and Coefficient of Filling for some Materials

Materials	Speed (V m/s)	Coefficient of filling ψ
Flour, formula feed	1-1.6	0.85
Food grain	2-3.2	0.75
Sawdust, woodchips	1.25 -2	0.8
Sand, soil, ash	1 – 2	0.8

Table 7.2: Bucket Modification for Difficult Materials

Materials	Modifications
Abrasive product	Front lip of bucket fortified to minimize wear when digging materials
Fluffy and powdery material susceptible to aeration	Vented buckets with holes help the product to settle
Watery and free-flowing	High point lip of bucket to provide maximum capacity
High density product	Strong buckets with heavy-duty belting or chain
Cohesive, wet or sticky products	Shallow rounded or flat bottom buckets not to allow materials to pack at corners

Equation 7.3 determines the shape of the bucket that will convey an average material (dry, free flowing, medium lump size <100 mm) at ambient to moderate temperature, slightly - moderately abrasive and not especially friable. Unusual or difficult materials necessitate appropriate modifications at the design stage as summarized in Table 7.2.

Since lump size, abrasiveness and cohesiveness of products are the principal factors to be considered when selecting bucket type and discharge option, Table 7.3 serves as a guide. However, the dimension of the buckets could be selected from Tables 7.4 and 7.5

Table 7.3: Recommendations for Selecting Bucket Elevators

Kind of load	Typical loads	Type of Elevator	Type of buckets	Coefficient of bucket filling ψ	Speed (m/s)	
					Belt	Chain
Dry powdered pulverised materials	Flour, feed, cement, phosphate fertilizer	High speed centrifugal or slow speed gravity	D or S	0.85	1.16	0.6-0.8
Granular, fine lump, low abrasive	Food grain, sawdust, wood chips,	High speed centrifugal discharge	D	0. 7-0.8	1.25-3.2	1.0-1.6
Ditto, highly abrasive	Gravel, ore, slags	slow speed, directed gravity	V or R	0.75 - 0.85	0.4-0.8	0.4-0.63
	Sand, ashes soil	High speed centrifugal	D	0.7-0.8	1.6-2.0	-
Sluggish, powdered, granular, moist	Wet sand or powdered chalk	Ditto	S	0.4-0.6	1.25-1.8	1.0-1.6
	Most chemicals, fluffy peat	Slow speed gravity discharge	S	0.4-0.6	-	0.6-0.8
Large lump, highly abrasive	Ore, palm fruit bunches	Ditto	V	0.6-0.8	-	0.5-0.8

*D = Deep, S = Shallow, V = V-shaped, R = Round bottom.

The buckets are shaped with either sharp or rounded bottoms to facilitate discharge. They are constructed from sheet steel (2-6 mm thickness), moulded from nylon or polypropylene or cast from malleable iron. They are fastened to the belt with small-diameter bolts (Fig.7.4) having flat heads or welded to a single or a double strand chain assembly (Fig.7.5). Chain speeds are limited to 1.3 m/s, while belt speeds can reach 3.2 m/s.

Table 7.4: Bucket main characteristics for belts

Bucket width B mm	Belt width B_b mm	Pitch of widely spaced buckets t_b mm	Deep Buckets D type		Shallow Buckets S type		Pitch of closely spaced buckets t_b mm	Buckets with side guides			
								V-type		R-type	
			i_o litre	i_o/t_b l/m	i_o litre	i_o/t_b l/m		i_o litre	i_o/t_b l/m	i_o litre	i_o/t_b l/m
100	125	200	0.2	1.0	0.1	0.50	-	-	-	-	-
125	160	220	0.4	1.3	0.2	0.66	-	-	-	-	-
160	200	320	0.6	2.0	0.35	1.17	160	0.65	4.06	-	-
200	250	400	1.3	3.24	0.75	1.57	200	1.30	6.50	-	-
250	300	400	2.0	5.0	1.40	3.50	200	2.0	10.0	-	-
320	370	500	4.0	8.0	2.70	5.40	250	4.0	16.0	6.4	25.6
400	450	500	6.3	12.6	4.20	8.40	320	7.8	24.4	14.0	43.7
500	550	630	12.0	19.0	6.80	10.80	400	-	-	28.0	70.0
650	700	630	16.8	26.6	11.5	18.20	500	-	-	60.0	120.0
800	-	-	-	-	-	-	630	-	-	118.0	187.0
1000	-	-	-	-	-	-	630	-	-	148.0	235.0

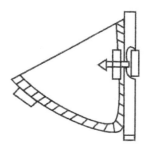

Fig. 7.4: Bucket fitted to belt

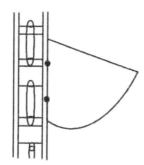

Fig. 7.5: Bucket welded to chain

Table 7.5: Bucket Main Characteristics for Chains

Characteristics	Width of bucket, B mm						
	400	500	650	800		1000	
Length of Bucket, L mm	470	595	595	595	740	740	920
Bucket pitch, t_b mm	500	630	630	630	800	800	1000
Geometrical volume of bucket, i mm³	30	60	75.5	96	149	186	288
Linear mass of moving parts, kg/m	120	150	180	194	220	270	300
Capacity (m³/h) at $V = 0.315\,m/s, \psi = 0.85$	65	103	124	165	200	250	310
Maximum size of lumps, mm							
Ungraded	160	200	220	220	250	250	300
Graded	100	125	160	160	200	200	300
Allowable tension in chain, kN	80	126	126	126	200	200	250
Highest lift of material (if $\varrho = 1\,t/m^3$)	53	54	45	38	49	36	50

7.5.2 Principal Criteria for Bucket Unloading/Discharge

During the ascending run (Fig.7.6a) the only force acting on the loaded bucket is the gravity force G (= mg). As the bucket turns round the head-wheel, the centrifugal force F (= mV²/r) also acts on the load. Their resultant R (Fig. 7.6a) changes in both magnitude and direction as the bucket moves round the head-wheel. Yet, its line of action always passes through P, the pole, located vertically above the head-wheel centre, O, at a distance h_p.

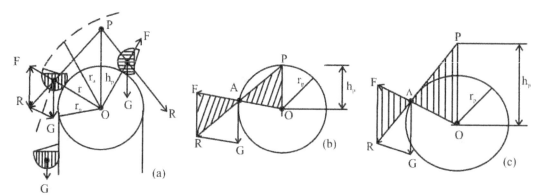

Fig 7.6: Discharge diagram of a bucket elevator

Note: M = mass of load in bucket (kg); V = velocity of centre of gravity A of bucket (m/s),
r = radius of rotation (m) of distance OA).
The gravity and acceleration (centrifugal) forces are given as Eqns. 7.4 and 7.5, respectively.

$$G = mg \qquad \qquad …7.4$$

$$F = mV^2/r \qquad \qquad …7.5$$

Using similarity of triangles AFR and AOP (Fig. 7.6a) give Eqns. 7.6 and 7.7, respectively.

$$h_p/r = G/F = mg/(mV^2/r) \qquad …7.6$$

$$h_p = g\,r^2/V^2 \qquad \text{(m)} \qquad …7.7$$

But $V = 2\pi rN$ *where* N *= rotational speed of head wheel (rpm) or*
$V = \pi rN/30$ *m/s* (Henderson and Perry, 1955)

Combining Eqns. 7.6 and 7.7 gives Eqn. 7.8.

$$h_p = gr^2 x30^2/\pi^2 r^2 N^2 = 895/N^2 \qquad \text{(m)} \qquad …7.8$$

The pole distance, h_p, depends solely on the rotational speed of the head pulley or sprocket. And the position of the pole point, P, can be used to classify the discharge characteristic of the bucket elevator. (Spivakovsky and Dyachkov, 1985).

When $h_p < r_p$ i.e. P is inside the pulley circle, $F >> G$ and the elevator is classified as centrifugal-discharge (Fig. 7.6b). When $h_p > r_p$ and P lies beyond r, $F << G$ and the elevator is classified as gravity-discharge (Fig.7.6c). *When* $r > h_p > r_p$ the discharge is by the combination of gravity and centrifugal effects.

114

It can be said that the method of bucket unloading is determined by the ratio of the pole distance and the pulley radius as given by Eqn. 7.9.

$$A = h_p/r_p \qquad\qquad \text{...7.9}$$

However, using the similarity of triangles gives Eqn. 7.10.

$$V/r = V_o/r_p \qquad\qquad \text{...7.10}$$

where V_o = Velocity of pulley (m/s); r_p = pulley radius (m).

And substituting in Eqn. 7.7 and Eqn. 7.9 we have Eqn. 7.11

$$A = g\, r_p/V_o^2 \qquad\qquad \text{...7.11}$$

Then, the pulley diameter is given by Eqn. 7.12.

$$\therefore Pulley\ diameter, D_p = 2AV_o^2/g \qquad \text{(m)} \qquad\qquad \text{...7.12}$$

From Eqn. 7.12, it can be said that the method of bucket unloading is determined not by the magnitude of the bucket's velocity but by the ratio of that velocity to the pulley diameter. Some recommended relationships are given in Table 7.6.

Table 7.6:   Recommended values of A and D_p

Type of speed	Type of unloading	A	D_p
High	Centrifugal	≤ 1	$\leq 0.204\ V_0^2$
High	Combined	1-1.4	$(0.205 - 0.286)\ V_0^2$
Moderate	Combined over back wall of buckets	1.5-3.0	$(0.306 - 0.612)\ V_0^2$
Low	Gravity	> 3.0	$0.6\ V_0^2$

7.5.3: Trajectory of Centrifugal Discharge.

Centrifugal discharge elevators operate at high speeds since they rely on centrifugal force for proper discharge. The operating speed is critical since the proper discharge pattern is a function of head-wheel diameter, belt or chain speed and bucket design. The material that leaves the bucket at the head-wheel follows a parabolic path (Millier, 1958) until deflected by impact with the casing or discharge chute. Thus, the trajectory of the material after it leaves the bucket is important to the proper design of the casing and the operation of the discharge chute (Beverly et al., 1983; Koster,

1985). The speed of the belt or chain must be held within close limits in order that the trajectory will fall within the specified region of the chute.

Using Fig.7.7a, the particle leaves the bucket when the radial component of gravity force equals the centrifugal force (assuming no sliding and no air resistance), given by Eqn. 7.13.

$$Mg\cos\theta = mV^2/r \qquad \qquad ...7.13$$

where θ is measured from the vertical

At point of discharge, the bucket is at position given by Eqn. 7.14.

$$\theta = \theta_a = \cos^{-1}(V^2/gr) \qquad \qquad ...7.14$$

and the particle leaves the bucket following velocity at angle θ_a (downward) to the horizontal (Fig.7.7b). The particle's position at time t is given by Eqns. 7.15 and 7.16.

$$X = Vt\cos\theta_a \qquad \qquad ...7.15$$

$$Y = -Vt\sin\theta_a - {}^1\!/_2\, gt^2 \qquad \qquad ...7.16$$

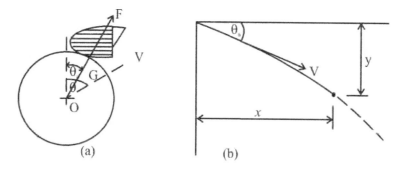

Fig. 7.7: Trajectory of particle discharged from an elevator bucket

By eliminating t and substituting for θ_a in Eqn. 7.14 gives Eqn. 7.17.

$$Y = -X\sqrt{(gr/V^2)^2 - 1} - X^2/2r\,(gr/V^2)^3 \qquad \qquad ...7.17$$

Equation 7.17 shows that the trajectory is parabolic and allows the position of the mouth of the discharge chute to be determined accurately so that the particles enter clearly into the casing without spilling outside.

It must be noted that when the loaded bucket is vertically over the head-pulley center O, F = G and R = 0. Even though there is no force on the material, discharge actually begins at this point given by Eqns. 7.18 and 7.19.

$$mg = mV^2/r \qquad \qquad \text{...7.18}$$

$$V^2 = rg \qquad \qquad \text{...7.19}$$

But $V = \pi r N/30$; $\therefore$ $\pi^2 r^2 N^2/900 = rg$ *and* $V/r = V_o/r_p$

Pulley rotational speed is given by Eqn. 7.20.

$$\therefore \quad N = \sqrt{(900g/\pi^2).(1/r_p)} \quad = \quad 29.91\sqrt{(1/r_p)} \qquad \qquad \text{...7.20}$$

where r_p = pulley radius (m), N = pulley rotational speed (rpm).

Equation 7.20 shows the relationship between the effective head wheel radius and its revolutions per minute for the most satisfactory discharge conditions.

7.5.4 Capacity of Bucket Elevators operating Vertically
This is one of the main parameters in the design of bucket elevators. The bucket elevator capacity is dependent on:

i. Volume of each bucket (litre)
ii. Bucket spacing (pitch) (m)
iii. Belt or chain speed (m/s)
iv. Weight of conveyed material (kg.)

The actual hourly volumetric capacity (or flow rate) is given by Eqn. 7.21.

$$V_s = 3.6\psi i_o V/t_b \qquad (\text{m}^3/\text{h}) \qquad \qquad \text{...7.21}$$

where i_o = struck volume of each bucket.

The mass throughput capacity is then given by Eqn. 7.22.

$$Q = \varrho V_s = 3.6\varrho\psi i_o V/t_b \quad (\text{t/h}) \qquad \qquad \text{...7.22}$$

Bucket spacing is governed by the shape of the bucket and its resulting discharge characteristics. They must be spaced to prevent discharge interference by the preceding bucket. Bucket spacing is from 2 to 3 times the projected width of the buckets. ψ (the filling coefficient) normally has a value of 0.65 - 0.75. But because the load that gets into the bucket depends on type of loading and pitch, shape and speed of buckets, ψ can vary from 0.4 to 0.8 (Table 7.3). Perry and Green (1984) have given some data on bucket elevator capacity depending on size of bucket, spacing, pulley diameter and belt speed. Some of these data for granular materials are given in Tables 7.7 and 7.8. The capacity of other products at the stated belt or chain speeds will vary in direct proportion to their bulk densities. The capacities and weights of V-bucket elevators are given in Table 7.9.

Table 7.7 Preliminary Selection Data for Centrifugal Discharge Bucket Elevators (Belt Type)

Size of Bucket (mm)			Bucket spacing (mm)	Pulley Diameter (mm)		Head pulley speed (rpm)	Belt speed (m/s)	Capacity (for ϱ = 1.6 t/m^3)	
Width	Projection	Depth		Head	Tail			kg/s	t/h
150	100	105	300	500	350	43	1.1	3.9	13
200	125	135	350	600	350	41	1.3	8.3	27
250	150	155	400	600	400	41	1.3	14.0	47
300	175	180	450	750	450	38	1.5	23.3	76
350	175	180	450	750	450	38	1.5	27.8	90
400	200	210	450	750	500	38	1.5	41.7	136

Table 7.8   Preliminary Selection Data for Continuous Bucket Elevator (Chain Type)

Size of bucket (mm)			Bucket spacing (mm)	Pulley Diameter (mm)		Head sprocket speed (rpm)	Chain speed (m/s)	Capacity (for ϱ = 1.6 t/m^3)	
Width	Projection	Depth		Head	Tail			kg/s	t/h
200	135	195	200	520	350	28	0.76	9.7	32
250	175	295	300	635	445	23	0.76	16.7	55
300	175	295	300	635	445	23	0.76	19.4	63
350	175	295	300	730	445	20	0.76	22.2	73
350	200	295	300	730	445	20	0.76	27.8	91
400	200	295	300	730	445	20	0.76	31.9	104
400	200	295	300	730	445	20	0.76	36.1	118

Table 7.9   Capacities and Weights of V-bucket Elevators

Buckets (mm)				Capacity (t/h) at 5.08 mm/s	Weight/metre of chain and bucket (kg/m)
Length	Width	Depth	Spacing		
304.8	304.8	152.4	457.2	26.4	53.6
406.4	304.8	152.4	457.2	29.1	59.5
508.0	381.0	203.2	609.6	39.1	81.9
609.6	508.0	254.0	609.6	91.0	96.7
762.0	508.0	254.0	609.6	114.7	104.2
914.4	609.6	304.8	762.0	156.5	139.9
1066.8	609.6	304.8	762.0	182.0	156.3
1219.2	609.6	304.8	914.4	174.7	223.2

7.5.5   Driving Power of Vertical Bucket Elevators

Since the bucket conveyor has its driving traction element as either belt or chain, the various resistance on the bucket elevator conveyor path is those given in the text dealing with belt or chain conveyors. The horsepower required to operate a bucket elevator is that required to scoop the material into buckets, to lift the material, to discharge the materials, to move a small amount of air, and to overcome friction in the bearings and other drive components.

However, for the bucket elevator, the main resistance can be summarized as the following: -

a. On the loaded side of the elevator, are the linear weights of chain or belt, bucket and load
    = L_1 (N/m)

b. On the return or empty side of the elevator, are the linear weights of chain or belt and bucket without load = L_2 (N/m)

c. At the boot, are the scooping resistance. This is difficult to determine mathematically. However, the work done can be expressed as an 'equivalent height' H_e (m) through which the bucket load must be carried.

The total circumferential force or equivalent tension of the belt or chain at the drive pulley is given by Eqn. 7.23.

$$T_e = (L_1 - L_2)((H + H_e)/t_b) \qquad \text{...7.23}$$

But $(L_1 - L_2) = \psi \varrho g i_o$ and H = difference in height between the feed and discharge points.

Hence, T_e can also be expressed as Eqn. 7.24.

$$T_e = (H + H_e)\psi \varrho g i_o/t_b = \left(1 + (H_e/H)\right)\psi \varrho g i_o/t_b \qquad \text{...7.24}$$

119

The total mass of load (G_n) on the up-going side of the elevator is the mass of one bucket times the number of buckets is given by Eqn. 7.25 and T_e given by Eqn. 7.26.

$$G_n = \psi \varrho g i_o H / t_b \qquad \qquad \ldots 7.25$$

$$\therefore T_e = G_n \left(1 + (H_e / H)\right) \qquad \qquad \ldots 7.26$$

H_e ranges from $3 - 12$ m depending upon the nature of the product and method of filling.

The frictional resistance and inertial effects as the loaded buckets revolve over the head pulley are small and are included in the overall drive efficiency, η ($= 0.80 - 0.85$). At initial start-up, considerable additional-power may be required so a safety factor is required as given by Eqn. 7.27.

$$HP = k T_e V / 1000 \eta \qquad \qquad \ldots 7.27$$

where $k = 1.15 - 2.0$ is the safety factor and V =belt or chain speed (m/s).

In summary, in the design of a bucket elevator for handling free flowing materials, one must determine (Theint and Lin, 2018): -

- The best bucket shape for pickup and discharge of material
- The belt or chain characteristics required
- The diameter of the head pulley
- The belt tension to provide the necessary lift at the correct speed
- The capacity and power requirements

7.6   MAINTENANCE OF BUCKET CONVEYOR

The main problem with bucket conveyors used for conveying very abrasive bulk material is wear at the feeding and discharge chutes, occasioned by sliding friction of the product flow. This problem is taken care of during design by using special abrasion-resistant steel or linings of polyurethane, rubber cushions, cover with basalt or aluminium oxide tiles.

Another problem is that of bent buckets which make excessive noise as they strike the elevator casing causing sparks and damaging the buckets. The remedy is to change deformed buckets as quickly as possible and check all bucket bolts for tightness. Also, to ensure that sheet metal edges of the casing do not protrude towards the belt.

The belts are always checked for excessive wear and replaced whenever damaged buckets or torn-out bolts are changed or when a weak point is noticed in them. For good preventive maintenance, the following are checked: -

1. The pulley for belt slippage, drive pulley lagging and shifting of the pulley on the shaft and the bearings.
2. The elevator for distortion of the elevator casing and frame by the movement of the supporting building or structure.
3. The drive mechanism (motor, gearbox, couplings, etc.) for alignment, greasing, lubrication and mounting.
4. The head casing for rubbing at the head shaft and loose sprockets on the head shaft.
5. The elevator boot for cleanliness.
6. All safety devices for proper functioning.
7. Worn/loose slide lagging; broken elevator buckets; excessive bucket and chain wear; loose fasteners.

When performing maintenance, understand and observe the following precautions: Do not replace or substitute bolts, nuts, or other hardware that is of lesser quality than the hardware supplied. Consult your dealer for proper replacements. Perform frequent inspections of the controls, safety devices, covers, guards, and equipment to ensure proper working order and correct positioning. After maintenance is completed, replace and secure all safety guards, safety devices, service doors, and cleanout covers. Do not climb ladder if damaged, wet, icy, greasy, or slippery. Maintain good balance by having at least two feet and one hand or two hands and one foot on ladder at all times. Use required safety harnesses and climbing equipment. Consult local safety authorities. Perform maintenance during normal daylight hours or in adequate ambient lighting (TRAMCO, 2014).

7.7 SOLVED PROBLEMS
Q 7.1 Calculate a vertical bucket elevator for the transportation of wheat into a silo.

Data: $Q = 80 \, t/h$; $H = 20 \, m$; $\varrho = 1/6 t/m^3$; $V_{belt} = 1.4 m/s$; $D_p = 700 \, mm$; $H_e = 8 \, m$; $\eta = 0.85$; $\psi = 0.8$; $k = 1.25$

Solution:

(a)   Main characteristics of the elevator

$D_p = 700$; $r_p = 0.35m$; $N = 60V/\pi D_p = 60x1.4/\pi x0.7 = \mathbf{38.2 \, rpm}$

Using Eqn. 7.8, $h_p = 895/N^2 = 895/38.2^2 = \mathbf{0.613 \, m}$

Since $h_p > r_p$ the buckets are gravity – discharged.

However, using Eqn. 7.3 the value of the struck ratio,

$i_o/t_b = Q/3.6V\varrho\psi = 80/3.6x1.4x1.6x0.8 = \mathbf{12.4 \, l/m}$

From Table 7.4, choose $i_o/t_b = 12.6$ t/m for deep type buckets with B = 400 mm,
t$_b$ = 500 mm, i$_o$ = 6.3 litre; B$_b$ = 450 mm. Also, the filling coefficient becomes:

$\therefore \; \psi = t_b Q/3.6V\varrho i_o = 0.5x80/3.6x1.4x1.6x6.3 = \mathbf{0.79}$

(b)   The total circumferential pull on the belt is got using Eqn. 7.25 and Eqn. 7.26,

$G_n = \psi\varrho g i_o H/t_b = 0.79x1.6x10^3 x9.81x6.3x10^{-3}x20/0.5 = \mathbf{3.125 \, kN}$

$T_e = G_n\big(1 + (H_e/H)\big) = 3.125(1 + 8/20) = \mathbf{4.375 \, kN}$

(c)   The driving power is got using Eqn. 7.27,

$HP = kT_e V/1000\eta = 1.25x4.375x1.4/0.85 = \mathbf{9.0 \, kW}$

Q.7.2: If the velocity of the c.g. of the bucket load is $1.24 \, m/s$ and its radius of rotation is 0.45m, calculate the pole distance. What is the method of unloading the bucket if the pulley radius is 0.30 m. For material of density $1.8 \, t/m^3$ and closely spaced buckets of depth 0.18m, find the bucket's useful volume if 120 t/h load is conveyed.   Take $V_{belt} =$

$1.35\ m/s\ and\ \psi = 0.85$. Neglecting scooping resistance, what driver power is required to lift the load 15 m. Take $k = 1.75$; $\eta = 0.80$

Data: $r = 0.45\ m$; $r_p = 0.30m$; $\varrho = 1.8t/m^3$; $t_b = h = 0.18m$; $Q = 120\ t/h$; $V_o = 1.24\ m/s$; $\psi = 0.85$; $H = 15m$; $k = 1.75$; $\eta = 0.80$; $V_{belt} = 1.35m/s$; $H_e = 0$

(a) Using Eqn. 7.8,

$V_o = \pi r N/30$; $N = 30V_o/\pi r = 30x1.24/\pi x0.45 = \textbf{26.3 rpm}$;

$h_p = 895/N^2 = \textbf{1.29 m}$

Since $h_p > r_p$ the buckets are gravity – discharged.

(b) Using Eqn. 7.3,

$i_o/t_b = Q/3.6V\varrho\psi$; $\therefore\ i_o = 120x0.18/3.6x1.35x1.8x0.85 = \textbf{2.9 l}$

Type of bucket is V-shaped.

(c) Using Eqn. 7.25 and Eqn. 7.26,

$H_e = 0$, Effective pull on the belt $T_e = G_n = \psi\varrho g i_o\ H/t_b$

$\therefore\ T_e = 0.85x1.8x9.81x2.9x15/0.18 = \textbf{3.627 kN}$

(d) Driving Power is got using Eqn. 7.27,

$HP = kT_eV/1000\eta = 1.75x3.627x1.35/0.8 = \textbf{10.7 kW}$

Q.7.3 A bucket elevator of pulley diameter 660 mm travelling at 2.8 m/s is to be used to unload wheat into a silo. Assuming the belt tensions are 3000 N and 1200 N respectively, show its method of unloading the wheat and calculate the motor power if $k = 1.2\ and\ \eta = 0.80$. Assume pulley belt resistance = 1.11.

Data: $D_p = 0.66\ m$; $V = 2.8\ m/s$; $T_1 = 3000\ N$; $T_2 = 1200\ N$; $k = 1.2$; $\xi = 1.11$; $\eta = 0.8$

Solution: $Using\ N = 30V/r_p\pi = 30\ x\ 2.8/0.33\ x\ \pi = \textbf{81.02 rpm}$

Using Eqn. 7.8, $h_p = 895/N^2 = 895/81.02^2 = 0.136\ m = \textbf{136 mm}$

Since $h_p < r_p i.e$ $136 < 330$ mm, the bucket is unloaded centrifugally.

$T_e = T_1 - T_2 = 3000 - 1200 = $ **1800 N**

The circumferential force on the pulley, $W = T_e \xi = 1.11x1800 = $ **1998 N**

Using Eqn. 7.27, $HP = kWV/1000\eta = 1.2 \ x \ 1998 \ x \ 2.8/1000 \ x \ 0.8 = $ **8.39 kW**

Q.7.4 Maize of density 1.48 t/m³ is conveyed in a closely spaced bucket conveyor of depth 0.75m. The speed of the c.g. of the bucket load = 0.80 m/s with a radius of rotation = 0.60m. If the pulley radius is 12 cm, what is the method of unloading? Find the bucket useful volume and the motor power given that $\varphi = 80\%$; $\eta = 90\%$; $V = 2.2 \ m/s$; $k = 1.15$; $W_t = 3.85 \ kN, Q = 12 \ t/h$.

Data: $V_o = 0.8 \ m/s$; $\eta = 90\%$; $\varrho = 1.48 \ t/m^3$; $\varphi = 80\%$; $r = 0.6 \ m$;
   $H = 0.75 \ m = t_b$; $k = 1.15$;
   $W_t = 3.85 \ kN$; $V = 2.2 \ m/s$; $r_p = 0.12 \ m$; $Q = 12 \ t/h$.

Solution: $V_o = rN\pi/30$; $N = 0.8 \ x \ 30/0.6 \ x \ \pi = $ **12.73 rpm**

Using Eqn. 7.8, $h_p = 895/N^2 = 895/12.73^2 = $ **5.5 m**

Since $h_p > r_p$, discharge is by gravity.

Using Eqn. 7.3, $i_o/t_b = Q/3.6V\varrho\psi$

$\therefore i_o = 12x0.75x100^3/3.6x2.2x1.48x0.8x1000 = $ **0.96 l**

Using Eqn. 7.27, the motor power $= kW_tV/\eta = 1.15x3.85x2.2/0.9 = $ **10.82 kW**

CHAPTER 7 Review Questions

7.1 Sketch a vertical bucket elevator and identify the constituent parts.

7.2 What are the advantages of belt-bucket over chain-bucket elevators?

7.3 Enumerate the advantages of bucket elevators.

7.4 Distinguish between different types of buckets used in bucket conveyors.

7.5 Describe briefly the methods of loading the buckets.

7.6 How are bucket conveyors unloaded? Derive from first principles the principal criteria for bucket unloading.

7.7 Describe the three principal variables in the design of bucket conveyors. What should the horsepower of a bucket conveyor be meant for?

7.8 On what components does the capacity of vertical elevators depend? Give the equations for the throughput capacity of a bucket elevator.

7.9 What forces or resistance must be overcome by a bucket elevator? What are the bucket elevator's horsepower required to do?

7.10 What would you check for good preventive maintenance of a bucket elevator? What are the major problems with bucket conveyors?

7.11 State the steps required in the design of bucket elevators for loading free-flowing materials.

7.12 For a bucket conveyor of $i_o = 2.5l$, $t_b = 450\ mm$; $r_p = 500\ mm$, $\psi = 0.75$ and $V = 2.5\ m/s$ carrying grain of $= 1.2\ t/m^3$, find the weight of bucket load and the throughput capacity. *(G = 2.25 kg; Q = 45.0 t/h).*

7.13 If the conveyor in Q. 12 above has r = 600 mm, what is the pole distance and how does it discharge? *(N = 39.840 rpm; h_p = 565 mm, combined discharge).*

7.14 What parameters influence the trajectory of centrifugal discharge? Show that the trajectory is parabolic. What will be the position of the load at $t = 3s$ if $\theta_a = 10^o$, $V = 1.8\ m/s$ and $r = 600mm$. *(x = 5.3 m; y = -45.1 m).*

8

SCREW (AUGER) CONVEYOR

8.1   INTRODUCTION

The screw conveyor has been found beneficial to agriculture for over 20 centuries now. The Archimedean screw was used for irrigation and lifting of sewage sludge. In the late 18th century, it was improved to transport grain and other dry bulk materials. The different configurations of the helical blade increase the applications of a screw conveyor for dry materials ranging from flour, powdered milk, ground feed, oil meals, fertilizer and food grains to peanut butter and palm fruit bunches.

Screw conveyors in modern industry are often used horizontally or at a slight incline as an efficient way to move semi-solid materials. These can have greater pitch spacing, resulting in a higher capacity without an increase in rotation speed. They usually consist of a trough containing either a spiral coiled around a shaft, driven at one end and held at the other, or a shaft-less spiral, driven at one end and free at the other. They are used for specific needs of transportation of bulk materials for shorter distances with lower transportation performance. These are conveyors without traction components (Tomašková and Sinay, 2014).

Screw conveyors are now being used as: -
- Batch or continuous mixers or blenders
- Feeders or unloaders (Puckett, 1958) where volume of material is controlled;
- Process involvement as in driers, hammer mills, oil expellers, cooler, cooker, de-watering unit
- Air or vapour lock devices.

They can handle damp, heavy viscous materials, hot loads and those chemically active which liberate offensive gases. The helical blades made of stainless steel, copper, brass, aluminium, cast iron, etc. can handle corrosive, or mildly abrasive materials. Typical applications are grain storage plants, feed mills, cereal processing plants and chemical plants. In general, they are used for pulverized or granular, non-corrosive, un-abrasive materials when the required capacity is moderate, when the distance is not more than about 61m, and when the path is not too steep. The high-speed auger is a very useful type of general-purpose conveyor for movement of grains at an angle. Trolley mounted shaft-driven augers are ideal equipment for filling isolated outdoor silos of medium height. Large-diameter augers fitted with a sweep collector assist in loading grains from the store into bulk transport vehicles.

Screw Conveyors can be operated with the flow of material inclined upward. When space allows, this is a very economical method of elevating and conveying. It is important to understand, however, that as the angle of inclination increases, the allowable capacity of a given unit rapidly decreases.

Thompson (1973) has summarized the advantages and disadvantages of screw conveyors thus:

Advantages

The screw conveyor has the following advantages:
- Low investment cost compared to other conveyors of similar capacity
- Compact design ensuring dust- or water-tightness
- Fairly simple fabrication with unsophisticated highly standardized equipment
- Lower maintenance and less moving parts than with most types of mechanical conveyors
- Ability to handle a wide range of solids

It usually costs substantially less than any other type of conveyor and readily made dust-tight by a simple cover plate (Altamuro and Hawkins ,1986).

Disadvantages

The disadvantages of a screw conveyor include:
- The use of lumpy, fibrous or sticky materials may cause problems
- The length is limited by the allowable torque capability of the drive and coupling shafts
- The power requirements can be high with solids that tend to pack
- The conveying efficiency is reduced when screws are inclined or vertical

8.2   TYPES OF SCREW CONVEYORS

There are three main types: (a) horizontal and slightly inclined (up to 20°), (b) vertically and steeply inclined and (c) conveying screw pipes. They may be used in conjunction with other types of conveyors in order to satisfy a particular need e.g. with endless belt, blowers, chain-and-slat, bucket elevators, etc.

(a)   Horizontal and Slightly-inclined

This widely used industrial screw conveyor consists essentially of a rotating screw formed by a helical blade welded to sections of a drive shaft which can then be coupled together to make up a conveyor of required overall length. The shaft is coupled to a drive and supported by outboard end bearings. For long conveyors one or more hanger bearings are provided at strategic locations along the shaft to prevent undue deflection (Fig.8.1). The fixed trough in which the shaft rotates is U-shaped, round-bottomed and is sometimes topped by a cover plate with an opening for loading and another at the other end bottom for unloading. The drive is connected at the discharge end. Though augers are usually driven by electric motor, or sometimes by internal combustion engine, they may also be driven by a hydraulic motor directly coupled to the shaft, and powered from a tractor's hydraulic system via flexible pipes. Tractors can operate 2-3 kW augers at 1100 – 1200 rpm for handling slurry or fertilizer, as well as grains.

(b)   **Vertically and Steeply-inclined**
     This is the enclosed screw or auger conveyor. The housing is a cylindrical tube. The screw is mounted in bearings at each end without intermittent hanger bearings (Fig.8.2). The lower end has a suitable hopper to give constant feed to the choke length (Roberts and Hayes, 1979).

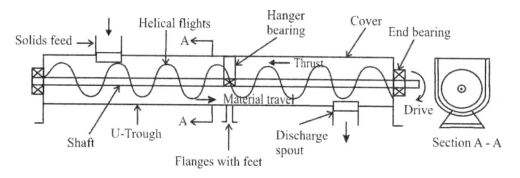

Fig. 8.1: Typical U-shaped horizontal screw conveyor

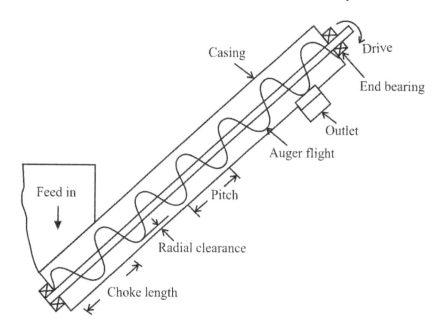

Fig. 8.2: Constructional features of an inclined auger conveyor

The drive motor is at the upper (discharge) end of the auger. The tubular casing ensures that the conveyor is kept full, prevents the fall back of material over the top and gives satisfactory operation. Auger conveyors are generally designed for relatively high speeds from 200 – 2000 rpm. When operated vertically, lift height can reach 15 m at a rate of 50 t/h but it must be supported by a screw feeder or a horizontal screw conveyor.

(c)   **Conveying Screw Pipes (Tubular Screw Conveyors)**
        These operate horizontally or at slight inclines up or down. For the conveying screw pipe, the helical blade is attached to the surface of the pipe. As the pipe rotates, the bulk material fed into it at one end is poured over helix edges, causing the load to advance with every revolution of the tube, and discharged at the other end, (Fig. 8.3). The helical blade width is usually (0.2 - 0.3) D where D = inner pipe diameter.

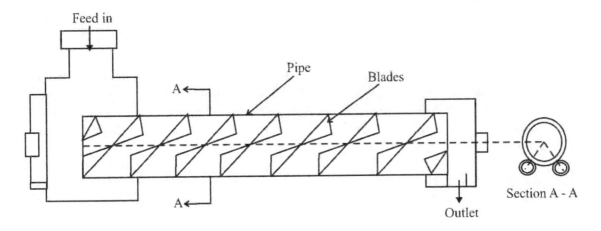

Fig 8.3: Tubular screw conveyor

8.3   TYPES OF FLIGHTS (CEMA, 1971; Colijn, 1985)

a.   The regular pattern is the helicoidal spiral for handling dry materials sizing from flour to medium size lumps.
b.   The ribbon pattern is used for conveying substances that are sticky, gummy or viscous.
c.   The cut flight with notches cut in the periphery of the helicoids is used for conveying, cooling and moderately mixing materials simultaneously.
d.   The cut and folded flight is for continual lifting and tumbling of the material, promoting agitation and aeration and improving mixing.
e.   The cut flight with paddles increases agitation and mixing.
f.   The regular flight with paddles is used to mix materials where the conveyor length provides time for proper mixing.
g.   The short pitch flight is used in inclined conveyors of > 20° and as feeder screws. They retard flushing of materials of a fluid nature.
h.   The conveyor screw paddles are used for mixing, blending, and stirring dry or fluid materials.
i.   The double flight conveyor screw is used to promote smooth and gentle flow of material.

129

j.   The stainless-steel conveyor screw is suited for sanitation in the conveying of foods, drugs, chemicals and allied products, for resistance to corrosion and for applications involving moderate to extreme heat.

k.   The tapering flight conveyor screw is used as feeder screws and for drawing materials from feed opening.

l.   The steeped diameter screw conveyors are used as feeder screws with the smaller diameter located under bins or hoppers to regulate the flow of material.

8.4   DESIGN CONSIDERATIONS FOR SCREW CONVEYORS

The process of screw conveying is a complex process with a great difficulty in developing analytical models to predict the design parameters of volumetric capacity and power requirement (Srivastava et al., 1993). A range of parameters, such as auger dimension, screw rotational speed, screw clearance, conveyor intake length and conveying angle for horizontal, inclined and vertical screw conveyors, pitch of auger flighting, and type of material being conveyed (Nicolai et al., 2006) are required for the design and performance of screw conveyors (Nicolai et al., 2004). This has led to the use of dimensional analysis for its performance prediction (Rehkugler and Boyd, 1962; Zareiforoush et al. 2010).

8.4.1   The Conveyed Product:

The screw conveyor tends to tumble and shear the conveyed material making it aerated and resulting in decreased bulk density. The more free-flowing the product, the less power required to transport it in the screw conveyor. A screw conveyor consists of a circular or U-shaped tube which a helix rotates. Grain is pushed along the bottom of the tube by the helix; thus, the tube does not fill completely (Olanrewaju et al., 2017, Peshatwar et al., 2020). Thus, the design mass throughput capacity should be based on the apparent bulk density. Sticky materials with long stringy particles are not suitable for screw conveying.

8.4.2   Screw Size and Rotational Speed:

Even though depth of loading is normally limited to a maximum of 45% of trough cross-section (Fig. 8.4), the choice of a suitable screw size depends on:

-   Overall diameter of conveyor
-   Diameter of shaft
-   Radial clearance (12-15 mm)
-   Type and pitch of flight
-   Proportion and size of lumps in the product being handled

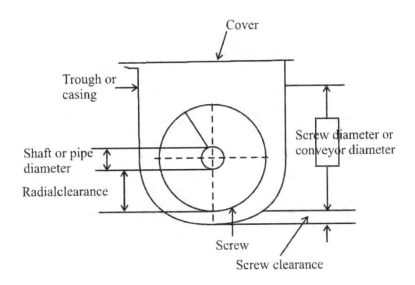

Fig. 8.4: Cross-section of U-trough screw conveyor

The ratio R = radial clearance/lump size and the percentage of lumps in the conveyed material have been used (Table 8.1) to determine the maximum lump size for different screw diameters. The values may be used as guides as the nature of the lumps (whether hard or degradable) may be relevant. Hard lumps must be provided for while degradable ones impose no limitations to the values in Table 8.1. The percentage lumps in the mixture of fines and lumps are the proportion of lumps ranging from maximum size to one-half maximum size.

The average velocity V (m/s) of the material as it is moved forward in the conveyor is given by Eqn. 8.1:

$$V = tN/60 \quad (m/s) \qquad \qquad \text{...8.1}$$

Table 8.1: Screw Conveyor Maximum Lump Size:

Screw diameter (mm)	10% Lumps R = 1.75 Max Lump (mm)	25% lumps R = 2.5 Max Lump (mm)	95% lumps R = 4.5 Max Lump (mm)
100	35	10	5
125	40	15	10
160	45	20	15
200	50	30	20
250	60	40	25
315	75	50	30
400	100	65	40
500	125	80	50
630	150	100	65
800	175	150	85

where t = pitch of the screw (m) and N = rev/min. But the rotational speed of the screw to be chosen is dependent on the kind of material conveyed and the screw diameter. The highest allowable rotational speed (rpm) is found by Eqn. 8.2.

$$N_{max} = A_c\sqrt{D} \quad (rpm) \qquad \qquad ...8.2$$

where D = screw diameter (m); A_c = constant that depends on nature of material (Table 8.2).

Also, screw diameters for conveying lumpy materials may be selected from Eqn. 8.3.

$$D \geq (10 - 12)a \quad or \quad D \geq (4 - 6)a_{max} \quad (m) \qquad \qquad ...8.3$$

where a = Lump size of graded (uniform) material (m)
a_{max} = Largest lump size in ungraded material (m)

8.4.3 Throughput Capacity:

The throughput capacity of a screw conveyor depends on the trough diameter (D_t) and shaft diameter (d) or screw diameter (D), the pitch of screw, $t = (0.5 - 1.0)D$ (m), rotational speed (N rpm) and the coefficient of filling ψ (Table 8.2) and inclination factor, K (Table 8.3).

Table 8.2: Values of Maximum Speed (N_{max}) and Coefficients ψ, A_c and C_o for Screw Conveyors.

Kind of Load	N_{max} rpm	Coefficients		
		ψ	A_c	C_o
1. Light, non-abrasive e.g. barley, corn meal, brown and clean rice, grain, flour, sawdust, Amaranth seeds, etc.	90-170	0.40-0.45	60-65	1.2
2. Light, very slightly abrasive e.g. cotton seed, soybean, corn, chalk, peat, cracked palm nuts, etc.	70-120	0.30-0.38	50-60	1.6
3. Heavy, low abrasive e.g. dry cotton seed, dry milk, salt, borax, asbestos, dry clay, etc.	60-95	0.25-0.31	45-50	2.5
4. Heavy, abrasive e.g. unshelled groundnut, unshelled castor bean, sand, cement, unshelled egusi-melon, un-milled rice, etc.	50-70	0.13-0.23	30-35	4

Table 8.3: Values of the Inclination Factor K for Inclined Screw Conveyors

Trough incline θ^o	0	5	10	15	20	25
K	1	0.9	0.8	0.7	0.65	0.4

The theoretical volumetric capacity (m^3/h) of an auger is given by Eqn. 8.4.

$$V_s = 3600AV \quad (m^3/h) \quad \quad \text{...8.4}$$

where A = cross-sectional area = $(\pi/4)(D_t^2 - d^2)$ or $(\pi/4)D^2$ (m^2); V = linear velocity of load (m/s). Substituting Eqns. 8.1 and A in Eqn. 8.4 gives Eqn. 8.4a.

$$V_s = 3600(\pi/4)(D_t^2 - d^2)tN/60 \quad (m^3/h) \quad \quad \text{...8.4a}$$

The part of A occupied by the load of density ϱ (t/m^3) introduces the coefficient of filling or loading factor, ψ or volumetric efficiency = V_a/V_s where V_a is the actual volumetric capacity
The throughput capacity of the screw conveyor by weight (t/h) is given by Eqn. 8.5.

$$Q = 3600\,\pi D^2/4\,.\,(tN/60).\,\varrho.\,\psi.\,K = 47D^2 t N \varrho \psi K \quad (t/h) \qquad \ldots 8.5$$

where K = inclination factor (Table 8.3) for designing inclined conveyors. When using Eqn. 8.5 for the design or selection of a screw conveyor, the following must be taken into consideration:

- ψ depends largely on load characteristics
- K depends on trough inclination
- N_{max} at which the device can be safely run is principally dependent on D, ψ and load characteristics and specified by the manufacturer.

Table 8.4 gives the values of N_{max} for the standard screw conveyor at different coefficients of filling. For other diameters of the screw, N_{max} can be obtained using Eqn. 8.6.

$For\ \psi = 0.45;\ N_{max} = 185.5 - 0.14D \quad (r = 0.991)$
$For\ \psi = 0.32;\ N_{max} = 130.2 - 0.12D \quad (r = 0.990)$
$For\ \psi = 0.15 - 0.30;\ N_{max} = 66.6 - 0.05D \quad (r = 0.968) \qquad \ldots 8.6$

N_{max} for different kinds of load are given in Table 8.2 and Table 8.4.

The value of ψ in Eqn. 8.5 should include corrections for thickness of flight and movement of material in the clearance space. Also, the special design of some screws (Section 8.3) will have some effect on the quantity Q for which capacity factors, C_1 (for special types of flights) and C_2 (for mixing paddles fitted within the flights) are introduced. These factors are given in Figs. 8.5 and 8.6 (Woodcock and Mason, 1987). The flight factor F_f used in determining the power of the screw conveyor is given in Table 8.5 and Eqn. 8.13. This factor is given for different flight types and coefficients of filling and ranged from 1.0 to 2.2. Bulk density of the conveyed material always plays an important role on the weight throughput capacity and invariably on the power of the conveyor. Some values for bulk density of certain agricultural materials are given in Table 8.6.

The equivalent capacity of a screw conveyor becomes (Eqn. 8.7 and Eqn. 8.7a).

$$Q = 47D^2 t N \varrho \psi K / C_1 C_2 \quad (t/h) \qquad \ldots 8.7$$

$$V_e = V_s x C_1 x C_2 \qquad \ldots 8.7a$$

for which the required speed for a given throughput capacity is given by Eqn. 8.8.

Table 8.4   **Standard Screw Conveyor Capacity (m³/h) and Maximum Speed (rpm)**

Screw diameter (mm)	Coefficient of filling							
	40 - 45%		30 -32%		25 - 30%		12.5 - 15%	
	N_{max}	V_s	N_{max}	V_s	N_{max}	V_s	N_{max}	V_s
150	165	10	120	5	60	3	60	1
230	155	40	100	15	55	8	55	4
305	145	80	90	33	50	18	50	9
357	140	124	85	50	50	29	50	15
406	130	172	80	71	45	40	45	20
457	120	230	75	96	45	57	45	29
508	110	292	70	124	40	71	40	35
610	100	464	65	201	40	123	40	62

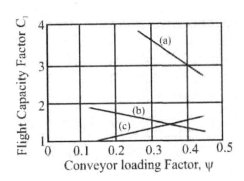 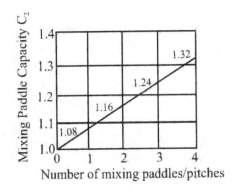

Fig. 8.5: Capacity factor C_1, for special flight screw conveyors (a) cut and foldedscrew flight; (b) cut flight; (c) ribbon flight; (C_1 = 1 for standard flight)

Fig. 8.6: Capacity factor C_2 for conveyors with 45°; reverse pitch mixing paddles fitted with the flights

135

Table 8.5: Flight Factor F_f

Flight Type	Per cent Conveyor Loading (filling)			
	15%	30%	45%	95%
Standard	1	1	1	1
Cut flight	1.1	1.5	1.2	1.3
Cut and folded flight	N.R*	1.5	1.7	2.2
Ribbon	1.05	1.14	1.2	-

*N.R = Not recommended.

$$N = (C_1 C_2 Q)/(47D^2 t \varrho \psi K) \quad (rpm) \qquad \dots 8.8$$

This required speed takes into consideration the capacity factors as well as the dimensions of the screw and the characteristics of the material to be conveyed.

Summary of Steps for Screw Conveyor Selection:
Primary considerations for the selection of a screw conveyor are: Type and condition of the bulk material to be conveyed including maximum particle size and specific bulk density. Capacity or feed rate of bulk material to be conveyed expressed in tonnes per hour, or cubic meter per hour.

i.   From the material to be conveyed, assess suitability for transport by screw conveyor.
ii.   If suitable, consider type of flight to use the loading factor (Table 8.5).
iii.   For lumpy materials, determine its size distribution and select minimum screw size from Table 8.1.
iv.   Determine maximum operating speed from Table 8.2 and Eqn. 8.6.
v.   Determine capacity factors C_1 and C_2 from Figs. 8.5 and 8.6 for the chosen screw type, and coefficient of filling.
vi.   Calculate Q at N_{max} from Eqn. 8.7. If this is greater than the required capacity use Eqn. 8.8 to determine necessary N; if higher capacity is desired choose a larger diameter screw and repeat from step iv.

Table 8.6   Material Factor F_m and Bulk Densities

Material	Bulk Density ϱ (t/m³)	F_m	Material	Bulk Density ϱ (t/m³)	F_m
Alumina	0.88-1.04	3.6	Oats (crushed)	0.30-0.38	1.2
Ammonium nitrate	0.72-0.99	2.6	Peas (dried)	0.72-0.80	1.0
Bentonite (fine)	0.8-0.96	1.4	Peanuts (un)	0.24-0.32	1.4
Bonemeal	0.8-0.96	3.4	PVC(Powders)	0.32-0.48	2.0
Cement (Portland)	1.51	2.8	PVC(Pellets)	0.32-0.48	1.2
Corn meal	0.64	0.8	Rice (clean)	0.72-0.77	0.8
China clay (Kaolin)	1.01	4.0	Rice	0.70-0.80	0.8
Cotton seed (dry)	0.40	1.8	Rye	0.70	0.8
Coffee (ground)	0.40	1.2	Sand	1.44-1.92	3.4-5.6
Cottonseed (hulls)	0.19	1.8	Sawdust (dry)	0.16-0.21	1.4
Flour (wheat)	0.53-0.64	1.2	Soap powder	0.24-0.80	1.8
Fly ash	0.48-0.72	4.0	Sugar (dry)	0.80-0.88	2.0-2.4
Gypsum (fine)	0.96-1.28	3.2	Talcum powder	0.80-0.96	1.6
Ice (crushed or cubes)	0.56-0.72	1.2	Wheat	0.72-0.77	0.8
Ice (flakes)	0.64-0.72	1.2	Wood (flour)	0.26-0.58	0.8
Lime (ground)	0.96	1.2	Wood (shavings)	0.13-0.26	3.0
Milk (powdered)	0.32-0.72	1.0	Beans (Soy)	0.72-0.80	1.0
Barley	0.61	0.8	Bran	0.26	0.8
Beans	0.77	0.8	Butter	0.95	0.8
Beans (Castor)	0.58	1.0	Corn (shelled)	0.72	0.8

Note: un = unshelled

8.4.4   Screw Conveyor Power:

The power requirement of a screw conveyor is a function of its length, elevation, type of hanger brackets, and type of flights, the viscosity or internal resistance of the material, the coefficient of friction of the material on the flights and housing, and the weight of the material. The power requirements of the screw conveyor increase with increase in the auger diameter because of the higher volume of the material conveyed with augers having larger diameters (Zareiforoush et al., 2010).

Until recently the power (kW) required to drive a horizontal screw conveyor is given by Eqn. 8.9.

$$HP_h = \ C_o QL/367 \ \ (kW) \qquad\qquad …8.9$$

and for sloping installations by Eqn. 8.10.

$$HP_s = (QH/367) + (C_oQL/367) \quad (kW) \qquad \ldots 8.10$$

where C_o = total friction factor (Table 8.2).

The horsepower requirement of a screw conveyor is very much dependent on the speed and increases almost linearly with the auger speed. Since the capacity of the inclined anger is much less than the horizontal, the power required is much greater because the power will also be used to raise the materials and overcome friction.

The use of overload or safety factor F_o is presently recommended to give the extra power required to start a full screw and to free a jammed screw and move a sticky material. No correction is necessary for values above 5 hp or 3.73 kW (Fig. 8.7).

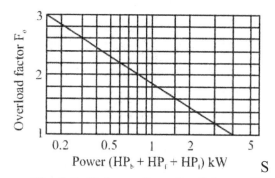

Fig. 8.7: Values of overload factor F_o

From Fig. 8.7, it will be observed that for

$hp < 0.746 \, kW, use \, F_o = 2$

$hp \, between \, 0.746 \, and \, 1.49 \, kW, use \, F_o = 1.5$

$hp \, between \, 1.49 \, and \, 2.98 \, kW, use \, F_o = 1.25$

$hp \, between \, 2.98 \, and \, 3.73 \, kW, use \, F_o = 1.1$

$hp \, over \, 3.73 \, kW, used \, F_o = 1.0$

To determine the normal input power, the drive efficiency η is introduced. Thus, giving the horsepower size of driving motor (HP_m) required as in Eqn. 8.11.

$$HP_m = HP_hF_o/\eta \quad or \quad HP_sF_o/\eta \quad (kW) \qquad \ldots 8.11$$

The current approach is to regard the total drive motor power as the sum of three power components, viz: (i) the power required to transport the bulk material horizontally at a uniform, regular specified feed rate in the screw conveyor (HP_h), (ii) the power required to overcome conveyor frictional resistance (HP_f), and the power to lift the material (HP_l), multiplied by the overload factor F_o and divided by the total drive efficiency η (0.85 − 0.95). This is given as in Eqn. 8.12.

$$HP_m = (HP_h + HP_f + HP_l)F_o/\eta \quad (kW) \qquad \ldots 8.12$$

The power to move the materials horizontally HP_h is given by Eqn. 8.13.

$$HP_h = F_f F_m QL/367 \quad (kW) \qquad \ldots 8.13$$

where F_f = flight factor (types of screw) (Table 8.5) and F_m = material factor (nature of material) (Table 8.6). This Eqn. 8.13 is similar to Eqn. 8.9 above with $C_o = F_f F_m$.

The power to overcome friction between moving parts is given by Eqn. 8.14.

$$HP_f = 75.7LND^{1.7} \quad (W) \qquad \ldots 8.14$$

while a simplified version of it is given by Eqn. 8.15.

$$HP_f = 50DL \quad (W) \qquad \ldots 8.15$$

Once the capacity of the inclined screw conveyor has been determined using Eqn. 8.7, the estimation of the additional power consumption, over that for horizontal operation is given by Eqn. 8.16.

$$HP_l = QH/367 \quad (kW) \qquad \ldots 8.16$$

where H is the vertical distance of the top end of the screw above the feed point with inclination of less than 20°.

The above approach and the discussion on the given factors are given by Carleton *et al.* (1969), CEMA BOOK NO. 350 (1971), British Standard BS 4409 Part 3 (1981) and FEM 2.121 (1985).

With the HP_m determined using Eqn. 8.12, the torque on the shaft or pipe needs to be calculated. The screw conveyor is limited in overall shaft or pipe length by the amount of torque that can be safely transmitted through the pipes and couplings which produce the speed. The torque (Nm) on the screw shaft is determined by Eqn. 8.17.

$$M_o = 30x10^3 HP_m\eta/\pi N \qquad \ldots 8.17$$

where HP_m = kW (as shown in Eqn. 8.12); η = drive efficiency and N = rotational speed (rpm).

The maximum admissible torques for a screw conveyor of different troughs and pipe diameters are given in Table 8.7.

The highest longitudinal force (N) acting on the screw is given by Eqn. 8.18.

$$F_l = M_o/r\ Tan\ (\alpha + \beta) = 2M_o/KD\ Tan\ (\alpha + \beta)\quad (N) \qquad \ldots 8.18$$

where r = radius of application of force P (m)

$\quad r = KD/2 = (0.7 - 0.8)D$

$\quad \alpha \quad = \quad$ angle rise of the helix line at radius r

$\quad \beta \quad = \quad$ angle of friction of material on screw surface, and

$\quad tan\ \beta \quad = \quad f = $ coefficient of friction.

Table 8.7 Maximum torque capacities for some trough and pipe diameters

Nominal diameter (mm)	Inside pipe diameter (mm)	Maximum admissible torque (Nm)
150, 230, 250	40	345.7
230, 250, 300	50	689.2
250	40	345.7
300, 360	60	1047.4
300, 360, 400	75	1853.0
450	90	1853.0
510, 610	100	2892.4

Having determined the throughput capacity, power requirement, torque and longitudinal force on the screw conveyor and its components, it is now easier to think about selecting the screw conveyor to do a particular job of conveying.

But there is no satisfactory method of selecting a screw conveyor as concise data are not available for individual design problems. The various effects of screw speed on auger capacity, volumetric efficiency and power requirement for various angles of auger inclination up to a certain speed make selection rather difficult. However, the following points should be borne in mind: -

- Capacity of screw conveyor increases as auger speed increases up to the point where it appears that centrifugal action of the flights on the material at the intake become so restrictive as to cause a decline in the capacity.
- The speed at which maximum capacity is obtained is a function of both the material and the screw conveyor size.
- The screw length has no effect on the capacity but capacity decreases as the angle of inclination increases.
- The capacity in the vertical position is from 30 to 40% the capacity in the horizontal position; this may be due to the restriction to grain flow in the intake of the conveyor at higher speeds and the fact that grain flows from a vertical orifice at

30% rate from a comparable horizontal orifice. In general, vertical screw conveyors operate at very high speeds.

- The horsepower requirement increases almost linearly with speed from 300 – 1000 rpm and varies with the type of material that is being elevated in the vertical conveyor. Increasing the angle of inclination causes the power to increase initially but a decrease follows beyond a certain angle due to the decline in the volumetric efficiency.

8.4.5   Vertical Screw Conveyors (Screw Elevators)

A vertical screw conveyor is one that conveys material upwards in a vertical path of more than 54° inclination. It consists of a conveyor screw rotating in a vertical tubular casing with a suitable inlet at the lower end and an outlet at the upper end.

The conveyor capacity is influenced by particle size, bulk density and particle shape as well as the clearance and the free length of the intake and the angle of repose of the product being conveyed.

The following special attributes make the vertical screw conveyor interesting:
- It is compact requiring less space than any other elevating conveyor.
- It does not move large lumpy, very dense or extremely abrasive materials.
- Materials are fed to the vertical by a straight or offset intake horizontal feeder screw conveyor to metre a controlled and uniform volume of material.
- The screw elevator runs full with discharge rate dependent on the speed of the screw feeder and not that of the screw elevator itself.
- There will always be material within the screw whether it is rotating or not; thus creating starting problems.
- Interruptions in loading result in stoppage of discharge as the elevator cannot empty itself.

The upward motion of bulk material along the vertical screw occurs by centrifugal and gravity forces against friction forces with lower axial speed (Fig. 8.8). However, the speed of the screw should not be less than the allowable (critical) lowest value which is required to move the material upwards along the screw.

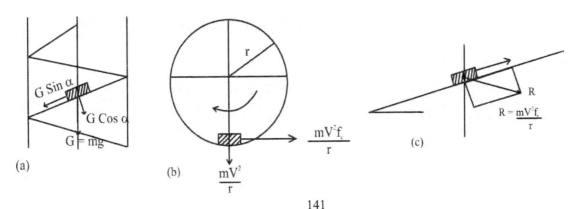

Fig. 8.8: Forces acting on a particle on the flight of a vertical screw conveyor

The sum of the horizontal forces along the screw (Fig. 8.8a) is given by Eqn. 8.19.

$$\sum F_x = G\,Sin\,\alpha + G f_c\,Cos\,\alpha \qquad \qquad …8.19$$

where f = coefficient of friction between material and conveyor surface.

As the particle moves along the screw round the shaft, it possesses centrifugal force, (F_c) (Fig. 8.8b) given by Eqn. 8.20.

$$F_c = mV^2/r. \qquad \qquad …8.20$$

There is a force of friction on the particle on the cylindrical wall of the casing, which acts in a horizontal plane (Figs. 8.8b and 8.8c) given by Eqn. 8.21.

$$F_w = mV^2 f_h/r \qquad \qquad …8.21$$

where f_h = coefficient of friction on housing (or casing) and
m = mass of particle

The component of this force on the helix surface is given by Eqn. 8.22.

$$F_s = (mV^2 f_h/r)Cos\,\alpha \qquad \qquad …8.22$$

and the force of friction on the helix surface is given by Eqn. 8.23.
$$F_p = (mV^2/r)f_c f_h\,Sin\,\alpha \qquad \qquad …8.23$$

There will be a critical velocity, V_{cr}, at which the particle will have no axial motion but will rotate together with the screw. V_{cr} is got from the Eqn. 8.24 to Eqn. 8.26.

$$mgSin\,\alpha + mgf_c Cos\,\alpha + (mV_{cr}^2/r)f_c f_h Sin\,\alpha - (mV_{cr}^2/r)f_h Cos\,\alpha = 0 \qquad …8.24$$

$$\therefore gr(Sin\,\alpha + f_c Cos\,\alpha) + V_{cr}^2 f_h(f_c Sin\,\alpha - Cos\,\alpha) = 0 \qquad …8.25$$
$$\therefore V_{cr} = \sqrt{(gr/f_h)((Sin\,\alpha + f_c Cos\,\alpha)/(Cos\,\alpha - f_c Sin\,\alpha))}$$

$$= \sqrt{(gr/f_h)Tan(\alpha + \beta)} \qquad \qquad …8.26$$

where $\beta = Tan^{-1}f_c$

But $V_{cr} = \pi r N_{cr}/30$

where N_{cr} = critical speed of rotation given by Eqn. 8.27.

$$\therefore N_{cr} = (30/\pi)\sqrt{(gr/f_h)Tan(\alpha + \beta)} = (30/\pi)\sqrt{(g/rf_h)Tan(\alpha + \beta)} \quad ...8.27$$

However, the vertical screw conveyor must not only convey the load, but overcome material fall back in annular clearance, impart linear velocity to the load against the pull of gravity. Thus, the actual screw speed $N > N_{cr}$ for the particles of the bulk material to move upward along the helix surface.

Since the vertical screw conveyor must run full, the actual throughput capacity is that of the screw feeder whose volumetric capacity assumes 95% cross-sectional loading for a flooded screw. Approximate capacities and recommended maximum rotational speeds of vertical screw conveyor are given in Table 8.8

Table 8.8 Vertical Screw Conveyor Capacities at Maximum Rotational Speeds.

Diameter of Conveyor Screw (mm)	Volumetric Capacity (m³/h)	Maximum rotational Speed (rpm)
152	10	400
228	35	300
304	75	250
406	170	200

The principles used to determine the power of the screw conveyor given in Eqn. 8.12 could be used to determine the power requirement in this vertical screw conveyor. However, actual power is determined experimentally since it is affected by many unpredictable variables.

8.4.6 Tubular Screw Conveyor (Conveying Screw Pipes)

A conveying screw pipe described in Section 8.2c has throughput capacity as given in Eqn. 8.5 or Eqn. 8.7 except that $t = 0.5D\ m, \psi$ is as in Table 8.9 and K is as given in Table 8.3.

Table 8.9 Values of the loading factor for ψ conveyor screw pipes.

Tube angle, θ	0	5	10	15	20
Ψ	0.22 - 0.33	0.19	0.13	0.10	0.08

Operations of this tubular screw conveyor cause continuous mixing action and should not be used for degradable materials. They are useful as processing equipment for calcination, drying and mixing materials. Tubular screw conveyors have the following advantages and disadvantages.

Advantages: -
- Simple construction
- Dependable operation
- Air-tight transportation

Disadvantages: -
- High dead weight
- High power consumption
- Bulkiness

For normal operation, the centrifugal force $(mr\omega^2)$ should be less than the gravitational force (mg) as shown in Eqn. 8.28.

$$\therefore \omega^2 r < g \qquad \qquad ...8.28$$

$But \; \omega = \pi N/30; \; g = 9.81 \, m/s^2 \; and \; r = D/2$

Therefore, the rotational speed is given by Eqn. 8.29.

$$\therefore N < 42.3\sqrt{(1/D)} \qquad \qquad ...8.29$$

To obtain the rotational speed of the tubular screw conveyor, consideration is given to the forces at play during its operation. The rotational speed of the tube should not be too high; otherwise, the material will rotate together with it, without the mixing action. In practice, the actual speed is given by Eqn. 8.30.

$$N = (20 - 30)\sqrt{(1/D)} \qquad \qquad ...8.30$$

For supporting the conveying tube on rollers as shown in Fig.8.9, the force of pressure is given by Eqn. 8.31.

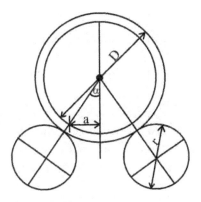

Fig. 8.9: Conveying tube on rollers

$$2P = (G + G_o)/Cos\ \alpha \quad (N/m^2) \qquad\qquad ...8.31$$

and Torque (Nm) on the pipe axis is given by Eqn. 8.32.

$$M = Ga + 2P[(2f_r + f_s d)/D_r]x\ D/2 \quad (Nm) \qquad ...8.32$$

where
D	=	diameter of pipe (m)
D_r	=	diameter of rollers (m)
d	=	diameter of roller axles (m)
f_s	=	coefficient of sliding friction of roller axles
f_r	=	coefficient of rolling friction of pipe rings
G	=	gravity force on load (N)
G_o	=	gravity force on pipe (N)
α	=	1/2 central angle of arranged support rollers (deg.)
α	=	0.25 D = distance between mass centre of material and central line (m).

The horsepower of the drive motor (kW) will be given by Eqn. 8.33.

$$HP = M\omega/1000\eta = (M\pi N/30).(1/1000\eta) = MN/9550\eta \qquad ...8.33$$

where ω = angular velocity (1/s); N = rotational speed (rpm); M = torque (Nm) and η = efficiency of drive transmission.

The drive motor power can also be derived using Eqn. 8.9 or Eqn. 8.10, but C_o is increased by 15 - 20% to allow for the high dead weight of the tube.

Summary

1. For screw conveyors, use U-shaped trough for horizontal conveying and tubular trough for vertical or inclined conveying.

2. A standard screw conveyor has pitch $t = D$; inclined screw conveyor ($< 20^\circ$), $t = 1/2D$, and steep or vertical screw conveyor $t =$ double, triple or variable pitch.

3. The actual volumetric capacity of a screw conveyor is much less than the theoretical volumetric capacity because linear motion was assumed instead of spiral motion and no friction factor. Also, there are clearances to allow the screw to rotate in the trough or tube.

4. The nature of the conveyed material is the most important design parameter for screw conveyors.

5. While the power requirement of screw conveyors increases almost linearly with speed (between 300 and 1000 rpm), throughput capacity decreases with speed.

6. The screw conveyor horsepower requirement is a function of screw length, elevation, types of hanger brackets, types of flights, viscosity and internal resistance of the material, coefficient of friction of the material on flights and housing, and the weight of the conveyed material.

7. The speed at which the screw conveyor maximum throughput capacity is obtained is a function of both the material and the screw conveyor size. Screw length has no effect on the throughput capacity but this capacity decreases as the angle of inclination increases. In general, vertical screw conveyors operate at very high rotational speeds.

8. The throughput capacity in a vertical position is from 30 to 40% times the throughput capacity in the horizontal position. This may be due to the restriction to grain flow in the intake of the conveyor at higher speeds and the fact that grain flows from a vertical onside at 30% rate from a comparable horizontal onside.

9. Throughput capacity of screw conveyor increases as auger speed increases up to a certain point after which it appears that centrifugal action of the flights on the material at the intake become as restrictive as to cause a decline in the capacity.

8.5 SOLVED PROBLEMS

Q.8.1: Select a horizontal standard pitch screw conveyor to transport $54\ t/h$ of soybean ($\varrho = 0.75\ t/m^3$). The conveyor should have a cut flight with one $45°$ reverse pitch mixing paddle per pitch, for a mixing time of 45 seconds minimum.

Data: $Q = 54\ t/h$, $\varrho = 0.75\ t/m^3$, material = soybean.
Conveyor; Cut-flight with 1 paddle/pitch, mixing time = 45 sec.

Solution:

$$V_s = Q/\varrho = 54/0.75 = 72\ m^3/h$$

For Soybean, $\psi = 0.32$; $F_m = 1.0$; $F_f = 1.15$; $C_1 = 1.57$; $C_2 = 1.08$; $C_o = 1.6$

Equivalent volumetric capacity (Eqn. 8.7a) $V_e = V_s x C_1 x C_2 = 72 x 1.57 x 1.08 = \mathbf{122.1\ m^3/h}$

From Table 8.4, a 508 mm diameter screw at maximum speed will do the job. The maximum speed = 70 rpm and $\psi = 30 - 32\%$. Pitch = diameter of screw = 0.508 m.

Length of the screw = N x length of one pitch x time (min.) = **70 x 0.508 x 45/60 m = 26.67 m.**

Overall mixing length of screw and trough length will be more than 26.7 m to provide space to bring the soybean into the trough and to discharge it from the trough without reducing the mixing time specified. Assume 4% extra length.

∴ Overall conveyor length = **1.04 x 26.67 = 27.74m.**

The power to drive the conveyor (Eqn. 8.9).

$$HP_h = C_o QL/367 = 1.6 x 54 x 26.67/367 = \mathbf{6.53\ kW}$$

$$HP_m = HP_h/\eta = 6.53/0.85 = \mathbf{7.68\ kW}$$

Under using current approach, put: $HP_m = (HP_h + HP_f)/\eta$

$$HP_h = F_f F_m QL/367 = 1.15 x 1.0 x 54 x 27.74/367 = \mathbf{4.694\ kW}$$

$$HP_f = 50DL/1000 = 50 x 0.508 x 27.74/1000 = \mathbf{0.705\ kW}$$

$$HP_m = (HP_h + HP_f)/\eta = (4.694 + 0.705)/0.85 = \mathbf{6.35\ kW}$$

Q.8.2: Find the throughput capacity and the drive power required by a standard pitch screw conveyor used for transporting bone-meal a distance of 20 m horizontally. What new power is required if it is conveyed the same distance up to a slope of 150? The initial data include:-
$\varrho = 0.90 t/m^3$; $t = D = 0.4\,m$; $\eta = 0.85$; $K = 0.7$; $For\ bone-meal\ \psi = 0.4$; $C_o = 1.2$; $N_{max} = 130\,rpm$; $F_m = 3.4$

Solution:
The mass throughput capacity (Eqn. 8.5) = $Q = 47D^2\ tN\ \varrho\ \psi$

$$Q = 47\ x\ 0.4^3\ x\ 130\ x\ 0.90\ x\ 0.4 = \mathbf{140.8\ t/h}$$

Volumetric throughput capacity = $V_s = Q/\varrho = 140.8/0.9 = 156.4\ m^3/h$

The horizontal drive power (Eqn. 8.9),

$$: HP_h = C_o QL/367 = 1.2x140.8x20/367 = \mathbf{9.21\ kW}$$

∴ the motor horsepower, taking into consideration the motor efficiency is (Eqn. 8.11).

$$HP_m = HP_h/\eta = 9.21/0.85 = \mathbf{10.83\ kW}$$

The height of lift, H $= \mathbf{20sin\ 150^o} = \mathbf{5.18\ m, and\ the\ slope\ factor, K = 0.7.}$

The total drive power up the incline is (Eqn. 8.10).

$$HP_s = K[(QH/367) + 9.21] = 0.7[(140.8x5.18/367) + 9.21] = \mathbf{7.84\ kW}$$

And the new motor power up the slope is (Eqn. 8.11).

$$HP_m = HP_s/\eta = 7.84/0.85 = \mathbf{9.22\ kW}$$

By going up to the incline, the slope factor reduces the throughput capacity of the conveyor and hence the power required to drive the motor. However, if the throughput capacity is to be maintained at 140.8 t/h, then the motor will be required to produce more power to drive it up the incline.

Q.8.3: For a vertical screw conveyor, $N = 1.5 N_{cr}$; $H = 10.8\,m$; $\eta = 0.90$; $\alpha = 300$; $\beta = 150$; $f_h = 0.28$. Find the drive power and the torque on the shaft when 114t/h of rice ($\varrho = 0.75\,t/m^3$) is being conveyed. Assume $F_o = 1.04$.

Solution:
The volumetric capacity is; $V_s = Q/\varrho = 114/0.75 = 152\,m^3/h$

From Table 8.7, a conveyor diameter of 0.406 m is chosen with maximum speed of 200 rpm. The critical velocity (Eqn. 8.27):

$$N_{cr} = (30/\pi)\sqrt{(2g/Df_h)tan(\alpha + \beta)} = (30/\pi)\sqrt{(2x9.81/0.406x0.28)tan(45)}$$

$$= \mathbf{125.45\,rpm}$$

$N = 1.5\,N_{cr} = 1.5\,x\,125.45 = \mathbf{188.2\,rpm}$.

This speed is adequate being less than the maximum. For the power (Eqn. 8.12),

$$HP_m = (HP_f + HP_l)F_o/\eta$$

For the friction power HP_f use Eqn. 8.14 and for horizontal power HP_l use Eqn. 8.16

$$HP_f = 75.7x10.8x188.2x0.406^{1.7}/60x1000 = \mathbf{0.554\,kW}$$

$$HP_l = QH/367 = 114x10.8/367 = \mathbf{3.355\,kW}$$

The required motor drive power: $HP_m = (0.554 + 3.355)/0.9 = \mathbf{4.517\,kW}$

The torque on the shaft is given by Eqn. 8.17 considering the drive efficiency and rotational speed

$$M_o = (30x10^3x4.517x0.9)/\pi x188.2 = \mathbf{206.27\,Nm}$$

Q.8.4 A screw conveyor is to be used to convey groundnut seeds of density $1.45\ t/m^3$ a horizontal distance of 25 m for processing. If the initial data given is: $D = 0.3\ m$; $N = 145\ rpm$; $t = 0.2\ m$; $\psi = 0.45$; $\eta = 0.9$; $\alpha = 15^0$; $A_c = 65$; $C_o = 1.2$, calculate the power required. If the conveyor is now inclined to a height $H = 4.2\ m$, find the additional power and the torque required. Assume $f_c = 0.44$; $N = 1.85 N_{cr}$; $k = $ slope factor $= 0.73$; where $\beta = Tan^{-1} f_c$.

Solution: Using Eqn. 8.5, $Q = 47 D^2 t N \psi \varrho k$ (but $k = 1$ for horizontal conveyor)

$Q = 47 x 0.3^2 x 0.2 x 145 x 0.45 x 1.45 = \mathbf{80.04\ t/h}$

Using Eqn. 8.9, $HP = C_o QL/367 = 1.2 x 80.04 x 25/367 = \mathbf{6.54\ kW}$

$\beta = Tan^{-1} f_c = Tan^{-1}(0.44) = \mathbf{23.75^0}$

Using Eqn. 8.27,

$N_{cr} = (30/\pi)\sqrt{(2g/Df_c)Tan(\alpha + \beta)} = (30/\pi)[(2x9.81/0.3x0.44)Tan(38.75)]^{\frac{1}{2}}$

$= \mathbf{104.31\ rpm}$

$Upwards\ speed\ N = 1.85\ N_{cr} = 1.85 x 104.31 = \mathbf{192.97\ rpm}$

$New\ Q_1 = 47 D^2 t N \psi \varrho k = 47 x 0.3^2 x 0.2 x 192.97 x 0.45 x 1.45 x 0.73 = \mathbf{77.76\ t/h}$
Using Eqn. 8.10,

$New\ HP_s = (QH/367) + (C_o QL/367) = (77.76 x 4.2/367) + (1.2 x 77.76 x 25/367)$

$= \mathbf{7.25\ kW}$

Using Eqn. 8.17,

$Torque = M_o = 30 x 10^3 HP_m \eta/\pi N = (30 x 1000 x 7.25 x 0.9)/192.97 x \pi = \mathbf{322.9\ Nm}$

$Additional\ power = 7.25 - 6.54 = \mathbf{0.71\ kW}$

Q.8.5: A horizontal screw conveyor with special cut flights and no paddles *has* $t = 0.5D, \psi = 0.25$ and runs at a maximum speed of 40 rpm. At what speed will it convey dry milk of *density* $0.55 \, t/m^3$ if $Q = 9.5 \, t/h$? What will be the power requirement if the milk is conveyed 40 m? Assume $\eta = 80\%$.

Data: $t = 0.5D$; $\psi = 0.25$; $N_{max} = 40 \, rpm$; $\varrho = 0.55 \, t/m^3$; $Q = 9.5 \, t/h$; $k = 1$; $L = 40 \, m$.

Solution: *From Fig.* 8.5, $C_1 = 1.65$; *Fig.* $8.6, C_2 = 1.0$; *Table* $8.2, C_o = 2.5$.

Using Eqn. 8.6, $N_{max} = 66.6 - 0.05D$, $\therefore 40 = 66.6 - 0.05D, \therefore \boldsymbol{D = 532mm}$.

$\therefore t = \boldsymbol{266mm}$.

Using Eqn. 8.8,

$N = (C_1 C_2 Q)/(47D^2 t \varrho \psi K) = (1.65x1.0x9.5)/(47x0.532^2 x0.266x0.55x0.25)$

$\therefore N = \boldsymbol{32.22 \, rpm}$

Using Eqn. 8.9,

$HP_h = C_o QL/367 = 2.5x9.5x40/367 \qquad = \boldsymbol{2.59 \, kW}$

Using Fig. 8.7, *assume safety factor* $F_o = 1.25$

Using Eqn. 8.11, $HP_m = HP_h F_o/\eta = 2.59x1.25/0.8 \qquad = \boldsymbol{4.05 \, kW}$

CHAPTER 8 Review Questions

8.1   Describe the ways screw conveyors are used in agricultural engineering practice. What materials can be conveyed by this conveyor?

8.2   What are the advantages and disadvantages of screw conveyors and tubular conveyors?

8.3   How many types of screw conveyors do you know? Describe one of them.

8.4   Name some types of flights of augers and what they are used for. Why is the actual volumetric capacity of tubular screw conveyors less than the theoretical?

8.5   Give the summary of steps for screw conveyor selection.

8.6   On what does the choice of a suitable screw size depend? What factors affect the throughput capacity of screw conveyors?

8.7   Develop the equation for the throughput capacity of a screw conveyor to involve capacity factor, filling factor, mixing factor, etc. What is the relationship between power requirement and throughput capacity of screw conveyors and their speed?

8.8   What does the following design parameters, ψ, K and N_{max}, of screw conveyors depend on? Why is the use of overload factor, F_o, recommended in screw conveyor design?

8.9   What makes up the total drive motor power of a screw conveyor? Give the equations of these different components. What are the power requirements of screw conveyors dependent on?

8.10   What are the special features of a vertical screw conveyor? What extra work does a vertical screw conveyor do?

8.11   Where are tubular screw conveyors used in agricultural engineering practice? What are its advantages and disadvantages?

8.12   Derive the critical speed of rotation of a vertical screw conveyor from first principles.

8.13   Design a screw conveyor for conveying palm fruits of density 1.9 t/m³ through a horizontal distance of 20 m. Initial data include: D = 0.25 m; N = 150 rpm; t = 0.22 m, ψ = 0.4, A = 65 and C_o = 12. If friction angle = 15° and helix rise angle = 30° at radius 0.8D/2, find the additional power required to convey the palm fruits over L = 20 m, and H = 5.4 m. Assume f_c = 0.4, N_1 = 2.25N_{cr}, η = 85% and slope factor = 0.7. Also, find the torque on the shaft. *(Q = 73.67 t/h; HP_{hor} = 4.82 kW; N_{cr} = 133.76 rpm; Q_1 = 103.47 t/h; HP_1 = 8.28 kW; ΔHP = 3.47 kW; M_o = 223.58 Nm)*

8.14   A tubular screw conveyor supported on rollers has the following data: G = 25 N; G_o = 60 N; α_o = 15°; a = 145 min; D = 560 mm; ψ = 0.3; θ = 10°; ϱ = 0.85 t/m³; K = 0.30; D_r = 600 mm; D_s = 80 mm; d = 20 mm; f = 0.025; K_1 = 0.175; N = 30 rpm; η = 0.90. Find the force pressure, the torque on the pipe axis, the drive motor power, throughput capacity and critical speed of the conveyor. *(P_f = 88 N; M_o = 389.2 Nm; HP_m = 1.36 kW; Q = 94.7 t/h; N_{cr} = 125.9 rpm)*

8.15   With D = 600 mm; t = 60 mm; ϱ = 0.90 t/m³; ψ = 0.45; Q = 100 t/h; K = 0.8; C_1 = 1.4; C_2 = 1.1; C_o = 1.6, find the required speed for the throughput capacity and the power required to lift grains up H = 5 m and L = 25 m; find also the torque on the drive staff. *(N = 468.2 rpm; HP_s = 12.26 kW; M_o = 250.1 Nm)*

8.16   For a vertical screw conveyor, m = 16 kg; V = 0.25 m/s; r = 150 mm; α = 15°; β = 10°; f_n = 0.35; f_c = 0.55, find the frictional forces, the critical velocity. *(F_c = 6.67 N;*

$F_w = 2.33\ N;\ F_s = 2.25\ N;\ F_p = 0.33\ N;\ V_{cr} = 1.4\ m/s;\ N_{cr} = 89.1\ rpm).$

8.17 If the mixing time in Q.16 above is 40sec, find the overall length of screw and the motor horsepower required to transport 60 t/h when $\eta = 0.85$, speed factor = 1.15; and 5% extra length for mixing. *($L_o = 21.5\ m;\ HP = 5.63\ kW;\ HP_m = 6.62\ kW$).*

9

SHAKING CONVEYORS

9.1 INTRODUCTION:

Shaking (Rocking) conveyors have found application in the industry for all types of bulk material handling (feeder or conveyor) of food products, chemicals, foundry sand, coal, iron ore, castings and scrap, and even in solids waste disposal (Dumbaugh,1985), under conditions at-times of gas-tightness. Inclined oscillatory conveyors are integral parts of cleaners, vibratory feeders, driers and spiral elevators.

They consist of a deck (trough or pan) supported by/on or suspended from/by a stationary frame (springs or hinged links). The deck is induced or caused to oscillate at some frequency with some amplitude by a powered crank. These parameters (Table 9.1) and the pattern of the motions of the deck and load determine whether the conveyor is an oscillating or a vibrating one. In the former, the bulk material slides along the deck exerting a variable or constant pressure on it and in the latter, conveying is carried out by causing the load to bounce along the deck with a throwing action. Wear due to sliding action limits oscillating conveyors to smooth granular or lumpy materials and make vibratory conveyors more popular today (Alexandrov, 1981).

Table 9.1: Operating ranges of frequency and amplitude for Shaking Conveyors

Type of Conveyor	Frequency (Hz)	Amplitude (mm)
Vibratory Feeder	13 – 60	12 - 1.0
Vibratory Conveyor	3 – 17	50 - 5.0
Oscillating/Reciprocating Conveyor	1 – 3	300 -50.0

The advantages and disadvantages of shaking conveyors are given below:

Advantages
The beneficial aspects of shaking conveyors include the following:
- Hot, poisonous, aggressive or abrasive as well as dirty and free-flowing materials can be handled.
- Drying, cooling, de-watering, mixing can be done while conveying.
- Scalping and screening or picking can also be done during conveying.

- A shaking conveyor system with divided flow stream or multiple discharge points in a unit is possible.
- Conveyor units are self-cleaning for good and proper sanitation.
- Conveyor units can be covered for air/gas/dust-tightness.
- The conveyor is simple in design and cheap in construction and with low headroom.
- The flow rate can be easily controlled by amplitude or frequency variation.
- A very wide range of conveyor sizes are available.
- The conveyor has low power consumption under steady state motion.
- There are no choking and no leakage.

Disadvantages

Some conditions do not favour the use of shaking conveyors. They include:
- Relatively short conveyor lengths (< 80m);
- Limited capacities (< 200t/h)
- Some degradation of material during operation.

9.2   OSCILLATING CONVEYOR:

An **oscillating/reciprocating** conveyor can be defined (Berry, 1958) as a trough or platform supported by oscillating arms which oscillates sinusoidally by the help of a crank in a direction lying in the vertical plane containing the longitudinal axis of the trough and at a tilt or link angle α to the vertical (Fig. 9.1). This conveyor operates by moving the whole carrying trough, and the material in it, forward and then leaving the material in the forward position by a rapid return stroke of the trough (Woodcock and Mason, 1987), with the direction of conveying depending on the direction or tilt from the crank arrangement (a small radius and long connecting rod). The velocity of the deck V_{deck} changes sinusoidally whereas it displaces rectilinearly perpendicular to the oscillating arm.

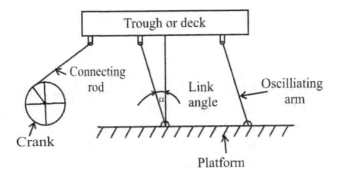

Fig. 9.1: Basic notation of a variable pressure oscillating conveyor

The forward motion of the conveyed material is by the sliding friction effect between it and the deck floor in spite of whether the vertical force exerted on the deck by the conveyed material is constant or variable. This conveyor is a horizontal trough varying in width from 305 to 1219 mm, mounted in rearward- inclined cantilever supports, and driven from an eccentric motor at 400 to 500 rpm. The effect is to "bounce" the material along at about 0.25 m/s with minimum wear on the trough. They have unit length of 30m and their capacities range up to 25 kg/s or 90 t/h with high efficiency and low maintenance.

An oscillating conveyor can be said to be completely specified by the:

- Link angle α to the vertical;
- Amplitude of oscillation (i.e. crank throw);
- **Frequency of oscillation**.

Berry (1958) suggested that for a rigid particle placed in the moving oscillating deck, there would be four different kinds of motion (Fig. 9.2) depending on the value of the deck frequency. If ω = frequency; f_s = coefficient of static friction; x_o , y_o = horizontal and vertical conveyor amplitudes, the equations of these types of equations are given below by Eqns. 9.1, 9.2, 9.3, 9.3a and 9.3b:

Type I: Very low frequency for which the particle will remain stationary on the deck with no slip whatsoever taking place.

$$0 < \omega^2 < (f_s g/(x_o + f_s y_o)) \qquad \ldots 9.1$$

Type II: Low frequency for which the particle must slip during part of the cycle and remains sticking to the deck over the rest cycle, (Stick-slip).

$$(f_s g/(x_o + f_s y_o)) < \omega^2 < (f_s g/(x_o - f_s y_o)) \qquad \ldots 9.2$$

Type III: Moderate frequency during which pure slip motion without loss of contact can exist. i.e. when downward vertical acceleration equals gravitational acceleration.

$$(f_s g/(x_o - f_s y_o)) < \omega^2 < (g/y_o) \qquad \ldots 9.3$$

Type IV: High frequency during which the particle is partly in contact with the deck and slipping, and partly off the deck surface and falling under gravity (bouncing). This type of motion has no equation, but using non-dimensional quantities, p' and q', Fig. 9.2. can be constructed to give the following expressions:

$$p' = f_s y_o/x_o \quad \ldots 9.3a; \qquad q' = x_o \omega^2/f_s g \qquad \ldots 9.3b$$

Most oscillating conveyors used in agriculture operate in the region of a pure slip, i.e. Type III. However, Eqn. 9.3 shows that the frequency can be increased without a limit, provided that y_0, the vertical amplitude of the deck, is correspondingly reduced.

9.2.1 Variable Pressure Oscillating Conveyor

Let us examine the relationships between the forces at play during a forward stroke and during a return stroke of an oscillating conveyor. With crank radius far smaller than the length of the connecting rod and that of the oscillating arm (Fig. 9.1), the deck velocity V_{deck} changes roughly as a sine wave (Fig. 9.3). Its acceleration j_{deck} can have both horizontal $j_{deck\,x}$ and vertical $j_{deck\,y}$ components.

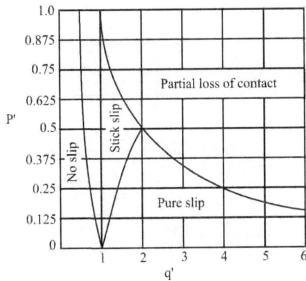

Fig. 9.2: Various modes of oscillating conveyor

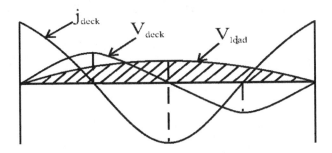

Fig. 9.3: Deck velocity and acceleration and load velocity curves for the conveyor in Fig. 9.1

Analysing the forces acting on the particle on the deck, it will be noted that $G = mg$ = weight of the particle, m = mass of the particle, f_s = sliding friction between the particle and the deck, r =

amplitude of oscillation or crank throw and N = rotating speed of the crank. The analysis of the forces at play is shown in Fig. 9.4 which also shows the directions of the deck accelerations.

During a forward stroke (Fig. 9.4a), the following accelerations and forces act on the particle:
i) $j_{deck\,y}$ is directed upwards;
ii) $mj_{deck\,y}$ presses load against the deck;
iii) $mj_{deck\,x}$ to displace the load forward along the deck, and
iv)   a friction force F_{fs} acts along the deck in the direction of stroke.

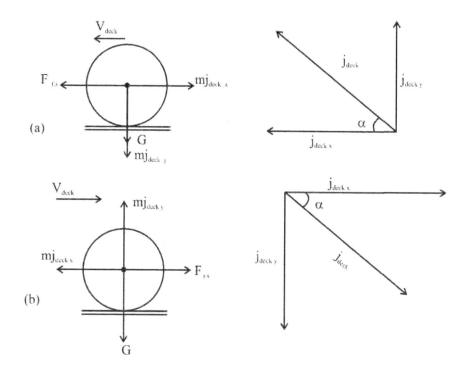

Fig. 9.4: Force diagram of an oscillating conveyor (a) forward stroke (b) return stroke

The load will continue its forward motion given Eqns. 9.4, 9.5, 9.6, and 9.7, if

$$F_{fs} < mj_{deck\,x} \qquad\qquad\qquad ...9.4$$

$$\therefore f_s m\left(g + j_{deck\,y}\right) < mj_{deck\,x} \qquad\qquad ...9.5$$

$$\text{But } j_{deck\,x} = j_{deck} Cos\,\alpha \quad and \quad j_{deck\,y} = j_{deck} Sin\,\alpha$$

$$\therefore j_{deck} > \frac{f_s g}{Cos\,\alpha - f_s Sin\,\alpha} \qquad\qquad ...9.6$$

$$\therefore j_{deck} = \frac{N^2 r}{90} > \frac{f_s g}{Cos\,\alpha - f_s Sin\,\alpha} \qquad \qquad \ldots 9.7$$

Equation 9.7 is the condition for the load to continue to slide forward during a forward stroke.

During return stroke (Fig. 9.4b), $mj_{deck\,y}$ is directed upward, thus reducing both pressure and friction. The load will continue its forward motion given Eqns. 9.8 and 9.9, if

$$mj_{deck\,x} > f_s m(g - j_{deck\,y}) \qquad \qquad \ldots 9.8$$

$$\therefore j_{deck} > f_s g / (Cos\,\alpha - f_s Sin\,\alpha) \qquad \qquad \ldots 9.9$$

Equation 9.9 is the condition for the load to continue to slide forward during a return stroke. However, for there to exist permanent load-to-deck contact during return stroke, the load pressure on the deck should be greater than zero, as given by Eqns. 9.10 and 9.11.

$$\therefore m(g - j_{deck\,y}) > 0 \qquad \qquad \ldots 9.10$$

$$or \quad j_{deck\,y} < g \qquad \qquad \ldots 9.11$$

i.e. vertical deck acceleration is less than the acceleration of free fall (gravity).

For crank-operated oscillating conveyors, the values of frequency are slightly greater than those given in Table 9.1 and those of amplitude slightly less ($r = 30 - 40\,mm, N = 300 - 400\,rpm$). In most oscillating conveyors used in agricultural engineering practice, the average travel speed of material $= 0.15 - 0.20\,m/s$, the depth of material in deck (trough) is $\leq 100\,mm$, the oscillating arm incline $\alpha = 20 - 30^o$ from vertical, opposite to load travel. For inclined conveyors, the arms should not be tilted by more than 15°, so as not to reduce the speed of upgrade travel of material.

The theoretical efficiencies are high at large link angles because the oscillatory motion is small compared with the steady forward velocity ($\eta = 0.80 - 0.98$). The horsepower for this type of conveyor is estimated from Eqn. 9.12.

$$HP = 4x10^{-4} QL / \eta Tan\,\alpha\,[(6x10^{-4}\,rN^2 / f_s) + 1] \qquad \qquad \ldots 9.12$$

9.2.2 Constant Pressure Oscillating Conveyor

The reciprocating action here is induced by a drag-link ABCO (Fig. 9.5) and the deck rests on rollers or balls. The crank OA rotates uniformly while crank BC does not, thus causing the connecting rod to oscillate. Condition for the load and the deck to move together is given by Eqn. 9.13,

$$j_{max} = Qgf_o/Q = gf_o \qquad \qquad ...9.13$$

As long as $j_i = j_{deck} \leq j_{max}$, the load will stick to the deck and move together with it. However, once $j_{deck} > j_{max}$, the load will slide along the deck due to inertia, given by Eqn. 9.14.

$$j_i = -Qgf/Q = -gf = constant \qquad \qquad ...9.14$$

Where: Q = weight of load on deck;
f_o = coefficient of sliding friction between the load and deck;
j_i = acceleration of the load;
f = coefficient of sliding friction in motion.

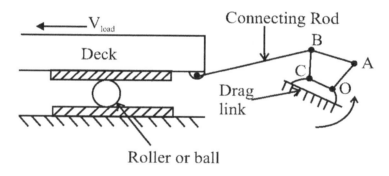

Fig. 9.5: Diagram of the constant pressure oscillating conveyor

From Fig. 9.6 we notice the motion of the load on the deck as the deck moves through one cycle, as follows:
i)   Load travels from O to A without sliding along the deck.
ii)   At A, $V_{deck,}$ changes abruptly and $j_{deck} = j_{max} = gf_o$ and the load starts sliding with some retardation until it reaches B.
iii) At B, $V_{deck,}$ changes direction and load slides further in the opposite direction until point C is reached.
iv)   At C, $V_{load} = V_{deck,}$ and the load travels on deck without sliding.

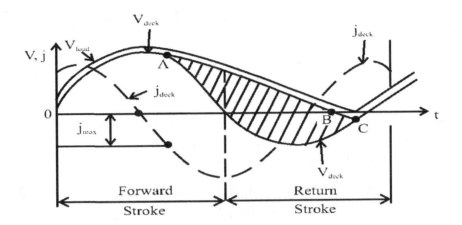

Fig. 9.6: Diagrams of load velocity, deck velocity and deck acceleration for constant pressure oscillating conveyor.

The distance covered by the load during one cycle of deck travel is the hatched area in the V-t curve (Fig. 9.6).

Advantages: Simplicity of construction; Low maintenance; Ease of adjustment for varying product flow; Few moving parts; Low power requirement; Gentle handling of fragile items; Versatility (perform special operations in addition to moving material); Silent operation with no transmission of annoying vibrations to operator; Capability of handling hot, corrosive, and dusty materials; Remote location of operating parts from discharge point; Abrupt discharge into space; Directional changes permitted with small change in elevation; Ease of cleaning (Schertz, 1962).

9.2.3 Oscillating Conveyor Calculations
This involves three interrelated problems namely:
1) Calculations of the oscillating system of the conveyor, i.e. amplitude and frequency.
2) Determination of the cross-sectional dimensions of the load-carrying element on the basis of the calculated average speed of the load and conveyor throughput capacity.
3) Determination of the power of the drive to overcome inertia forces, resistance to load motion and losses in the elastic elements of the oscillating system.

Coefficient of operating mode of the oscillating conveyor is given by Eqn. 9.15.

$$K_v = \text{maximum normal component of deck acceleration/} \\ \text{component of acceleration due to gravity}$$

$$= r\omega^2 Sin\,\beta/gCos\,\alpha \qquad\qquad\qquad ...9.15$$

where β is the angle of drive; α = link/inclination angle

161

Table 9.2: Recommended Coefficients of Operating Mode K_v

Type of Conveyor	Type of Vibrational Drive	K_v for Loads	
		Dusty & pulverized loads	Granular & lumpy loads
Single-deck or pipe, light and medium duty; Q ≤ 50t/h	Centrifugal or electro-magnetic	3.0 - 3.3	2.8 - 3.0
Single-deck or pipe, light and heavy duty; Q ≥ 50t/h	Centrifugal or electro-magnetic	2.0 - 2.5	1.8 -2.3
Single and double-pipe, balanced, light & medium duty; Q ≤ 50t/h; L≤ 30m	Eccentric	1.6 - 2.8	1.5 - 2.5
Single and double-pipe, balanced, heavy duty; Q ≥ 50t/h and L ≥ 30m	Eccentric	1.3 - 2.5	1.2 - 2.0

For horizontal conveyors, Cos α = 1, making Eqn. 9.15 giving Eqn. 9.16.

$$K_v = r\omega^2 Sin\,\beta / g \qquad \qquad \ldots 9.16$$

With $K_v < 1$, load is always in contact with the oscillating plane and moves without rising above it (rocking conveyors); with $K_v > 1$, load at some moments rises or jumps above the oscillating plane. Table 9.2 gives the recommended values of K_v.

The direction angle of oscillation, β (drive angle) is taken depending on the oscillation frequency:

with
$$\omega \geq 15\,Hz;\ \beta = 20 - 25^o$$
$$\omega < 15\,Hz\ \beta = 30 - 35^o$$
$$average\ \beta = 30^o.$$

The recommended amplitudes and frequencies of some drive types are given in Table 9.3.

The speed of the load (V m/s) is given by Eqn. 9.17.

$$V = (K_1 \mp K_2 Sin\,\alpha) r\omega Cos\,\beta \sqrt{(1 - 1/K_v^2)} \qquad \qquad \ldots 9.17$$

where K_1 and K_2 are empirical constants (Table 9.4) and the minus sign for upgrade conveyors.

Table 9.3: Recommended Amplitude (r) and Frequencies (ω) of Vibrations of Shaking Conveyors

Type of drive	Frequency (Hz)	Amplitudes (r, mm) for loads	
		dusty & pulverized	granular & lumpy
Electromagnetic	50	1.2 - 2.0	0.75 -1.0
Centrifugal Single	45 - 25	1.2 - 3.0	0.80-2.5
Centrifugal double	25 - 15	2.0 - 4.0	2.0- 3.0
Eccentric	13 - 7.5	5.0 - 15.0	4.0-8.0

Table 9.4: Coefficients K_1 and K_2 in Load Speed Equation 9.17

Kind of Load	Particle Size, mm	Moisture Content %	K_1	K_2
Lumpy	5 - 200	-	0.9 - 1.1	1.5 - 2.0
Granular	0.5 - 5	0.5 - 10	0.8 - 1.0	1.6 - 2.5
Pulverized	0.1 - 0.5	0.5 - 5	0.4 - 0.5	1.8 - 3.0
Dusty	up to 0.1	0.5 - 5	0.2 - 0.5	2 - 5

For horizontal conveyors, $\sin \alpha = 0$, and given by Eqn. 9.18.

$$K_1 r \omega \cos \beta \sqrt{(1 - 1/K_v^2)} \qquad \text{...9.18}$$

The throughput capacity is given by Eqn. 9.19.

$$Q = 3600 A \psi V \varrho_{loose} \quad (t/h) \qquad \text{...9.19}$$

where $A = Bh$; $h = 50 - 100\ mm$; $V = 0.2\ m/s\ maximum$; $\varrho_{loose} = 0.8\varrho$; $N = 50 - 100\ rpm$; $r = 50 - 150\ mm$; $\omega = 2\pi N$ $\psi = capacity\ factor = $ (a) $0.6 - 0.9\ for\ open\ decks$; (b) $0.6 - 0.8\ for\ rectangular\ pipes$; (c) $0.5 - 0.6\ for\ circular\ pipes$.

If it is possible to obtain the aggregate weight of load and moving conveyor parts (Q_o) in kg, then the horsepower required to drive this conveyor is simply got from Eqn. 9.20.

$$HP = 0.0014 Q_o \quad (kW) \qquad \text{...9.20}$$

More accurate methods of calculating power would require plotting the load vs. power curves for an operating cycle of the conveyor.

Because of the sliding action of the load on the deck and the rapid wear encountered in the process, the oscillating conveyor is being replaced by other types. Another disadvantage of this conveyor is that it is an unbalanced conveyor and is subjected to high impact loads.

9.3 VIBRATING/VIBRATORY CONVEYORS

These are shaking conveyors that throw particles from the conveying surface during transport. Transported load literally "bounce" along the deck in an essentially horizontal position due to vibration. Thus, the particle-to-particle cohesion must be greater than particle-to-deck adhesion. Vibratory movement is one of the most efficient methods of conveyance of granular and particulate materials. Vibratory conveyors (VC) are widely applied in many technological processes involving gravimetric transport, processing, and dosing of granular materials. From the macroscopic point of view, the process of vibratory conveyance is base d on recurrent micro-displacements of particles of the material being conveyed (Despotović and Stojiljkovic, 2005).

Basic Types of Operations in Vibratory Conveyors
- conveying along to the horizontal surface
- conveying along to the lightly inclined surface
- conveying along to the spiral elevator
- vibratory dosing; loading and (or) discharging
- vibratory finishing; pulverization and separation
- vibratory compaction
- vibratory rapping (electrostatic precipitators) (Despotović et al., 2015)

While in the oscillating conveyor the deck is supported by a solid/rigid arm, in the vibratory conveyor the deck is either suspended from, or supported by springs (Fig. 9.7) and connects to an exciter or vibrator which induces it to vibrate at a small amplitude (1-50 mm) but at high frequency (*up to* 3000 *rpm or* 50 *Hz*). The conveyor in Fig. 9.7 is directionally constrained to move perpendicular to the fixed guide springs at drive angle ß to the horizontal of about $20 - 30^o$. The conveyor operates at accelerations $j_{deck} > g$ causing the particle to travel forward in a series of short hops without sliding. These short hops are imperceptible to the eye (see Fig. 9.8b)

The chief assets of the vibrating conveyor are: -

– Ability to move loads under the conditions of complete air-tightness, freedom from contact between the materials conveyed and the moving parts of the conveyor.
– Ability to combine the function of transportation with the performance of necessary processing operations (drying, cooling, mixing, etc.).
– Low wear on the deck even if the load is abrasive because of the hopping/bouncing action of the load on the deck.
– Simplicity of construction, the possibility of being loaded and unloaded at any desirable point.

- Low power consumption under conditions of steady state motion.
- Suitable for carrying materials at a rate of up to 200 t/h over a distance of up to 80 m because of high conveyor acceleration $j_{deck} > g$.

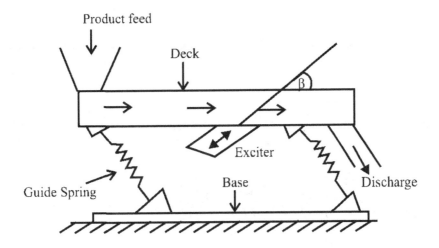

Fig. 9.7: Schematic diagram of vibratory conveyor

The disadvantages: -

It should be noted that vibrating conveyors
- Rarely stand to the rigours of service for longer than one year.
- Possibly transmit vibration to the structure on which they are built.
- Are essentially operated horizontally, and upgrades diminish its capacity at a rate of 3 to 5% per degree.

9.3.1 Principle of Bouncing Operation of Vibratory Conveyor

Once j_{deck} exceeds g the material will lift off the deck and follow a trajectory path (Fig. 9.8a). As the material is in flight, the deck continues its downward backward motion before it meets the particle at the next impact point during the next forward motion of the deck. At optimum operating conditions for maximum efficiency the impact point will occur just before the next lift-off point, allowing for a brief contact time and negligible abrasive wear. Many analytical studies have been done on the type of motion of particles on vibratory conveyors (Gutman, 1968; Ochmen, 1981; Ng et al.; 1982). These studies are rigorous and outside the scope of this book. However, suffice it to note that the complex movement of particles along the deck of a vibratory conveyor is influenced by many factors given in Table 9.5. Engineers now face the problem of determining a suitable combination of parameters to give the maximum transport rate of a specified bulk material along the deck.

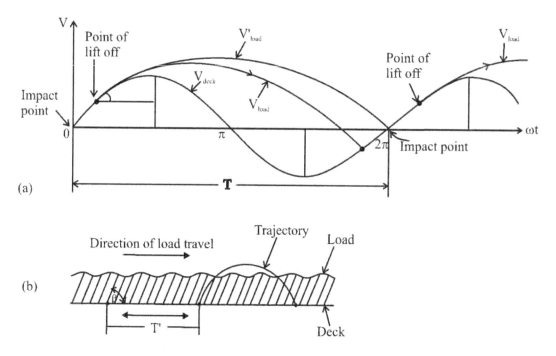

Fig. 9.8: Operating principle of a vibratory conveyor

9.3.2 Some Important Parameters for Calculating Transport Rate of Vibratory Conveyors

The following assumptions must be made in order to simplify our understanding of the transportation of materials in a vibratory conveyor: -

- Bulk materials can be represented as single-point-mass
- Impact between material and deck is non-elastic
- Air resistance is neglected
- Deck path is straight
- Rotation of materials is neglected
- Influence of deck walls are neglected
- Coefficients of static and dynamic friction are the same.

With the above assumptions, the following parameters are considered when calculating transport rate of vibratory conveyors:

a) **Throw factor or dynamic material factor** is the ratio of the vertical acceleration of the deck to the gravitational acceleration g given by Eqn. 9.21.

$$K_v = \omega^2 r \, Sin \, \beta / g = r(2\pi N)^2 / g \, Sin \, \beta = j_{deck \, y} / g \qquad \ldots 9.21$$

It is the factor which determines when the flight or throw phase begins, i.e. when $K_v \geq 1$. If $0 < K_v < 1$, then the material never leaves the deck. The other phases are the sliding phase and the impact phase and the rest phase occurring at different time periods.

166

Table 9.5: Factors Influencing Materials Movement on a Vibrating Conveyor

Deck Related		Material Related	
1	Type of motion & drive angle	1	Bulk density, particle shape & size distribution.
2	Frequency, amplitude and inclination angle	2	Internal & sliding friction coefficients.
3	Secondary vibrations in deck	3	Cohesion, damping, bed thickness and permeability.
4	Shape and smoothness of the deck and its liner.		

However, the most efficient operation is the one when the impact phase coincides with the next flight phase. At this instant Eqn. 9.22 obtains.

$$K_v = \sqrt{1 + (Cos\ 2\pi n + 2(\pi n)^2 - 1/2\pi n - Sin\ (2\pi n))^2} \qquad ...9.22$$

$where\ n = Time\ of\ flight(T_f)/Period\ of\ vibration(T_T) = flight\ ratio$

For design purposes, K_v = 1.5 - 2.0 even though the maximum is 3.3. For these design values, β ranges between $30 - 50^o$ (Colijn, 1985).

b) **Dynamic Machine Coefficient or Machine Number** is the ratio of the maximum deck acceleration to the gravitational acceleration g is given by Eqn. 9.23.

$K = maximum\ deck\ acceleration/gravitational\ acceleration$

$$= j_{max}/g = K_v/Sin\ \beta \qquad ...9.23$$

The relationship between amplitude and frequency is given as Eqn. 9.24.

$$r = j_{max}/\omega^2 = j_{max}/(2\pi N)^2 = gK/(2\pi N)^2 \qquad ...9.24$$

showing that amplitude decreases as frequency is increased. Thus, Eqn. 9.25 is obtained.

$$K = r(2\pi N)^2/g = K_v/Sin\ \beta \qquad ...9.25$$

Practical vibrating conveyors have K = 1 to 4 showing low frequency and high amplitude combination for fairly long pieces of equipment. Small rugged feeders with great structural integrity have K up to 12 and can withstand high frequency vibrations.

c)   **Velocity efficiency** (Eqn. 9.26) is the ratio of the average velocity of material to the horizontal component of deck velocity: -

$$\eta_v = V_{ave}/\omega r Cos\,\beta = V_{ave}/2\pi r N Cos\,\beta = V_{ave}/V_{deck\,x} \qquad \ldots 9.26$$

This efficiency of transport is a function of K_v, β and μ_f (coefficient of friction between material and deck surface) which can be related to two dimensionless numbers, K_v and $\mu_f\,tan\,\beta$ (Fig. 9.9).

Nonetheless, it also depends on some other parameters which are usually accounted for by empirical correction factors.

f_m *for material characteristics* $(0.8 - 0.9\ for\ heavy, granular, dry\ material)$.
f_h *for depth of material on deck* $(1.0\ for\ small\ depths\ to\ 0.75\ for\ depths\ of\ 300mm)$.
f_s *for slope of deck* $(1.0 - 0.1\ for\ 0 - 15^o\ up\ slope;\ 1.8\ for\ up\ to\ 15^o\ down\ slope)$.

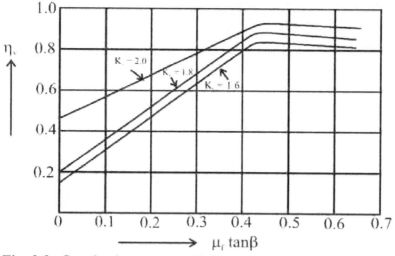

Fig. 9.9: *Graph of transport efficiency as a function of* K_v *and* $\mu_f tan\beta$

From the above analysis, we note the following given by Eqns. 9.27 and 9.28.

$$V_{max} = \omega r Cos\,\beta = 2\pi r N Cos\,\beta \qquad \ldots 9.27$$

We get η_v from Fig. 9.9, then

$$V_{ave} = \eta_v V_{max} = 2\pi\eta_v rN\cos\beta \qquad \qquad \ldots 9.28$$

∴ Transport velocity of the material along the deck will be got by multiplying Eqn. 9.28 by the correction factors given above to give Eqn. 9.29.

$$V_{actual} = \eta_v f_m f_h f_s r(2\pi N)\cos\beta \qquad \qquad \ldots 9.29$$

9.3.3 Throughput Capacity (Flow Rate) of Vibrating Conveyors
The volumetric flow rate equals the product of the actual velocity (Eqn. 9.29) and the material cross section area given by Eqn. 9.30.

$$V_s = V_{actual} A \qquad \qquad \ldots 9.30$$

Multiplying Eqn. 9.30 by the material bulk density gives the mass flow rate as Eqn. 9.31.

$$\therefore Q = \varrho A V_{actual} \qquad \qquad \ldots 9.31$$

9.3.4 Some Factors Influencing Design Parameters of Vibrating Conveyors
These are factors that need to be taken into **consideration** when calculating some **design** parameters of vibratory conveyors.
- Feed rates are sensitive to particle size distribution and moisture content.
- High conveying rates require low operating frequency but constant deck acceleration require high amplitudes of vibration.
- Increase deck acceleration up to a point whereby $j_{deck} > g$ slightly; excess j_{deck} results in irregular bouncing of material on the deck.
- Optimum angle of inclination, β, depends on both j_{deck} and μ_f. increasing material adherence to deck floor means use of smaller β.
- Lining deck floor with high-friction material (e.g. rubber) improves performance.
- Increasing depth of material on deck increases friction and improves conveying velocity.

Some reasonable average values of size, depth and conveying speed for some agricultural materials are given in Table 9.6 for design and capacity computations.

9.4 DRIVE MECHANISMS

There are essentially four main types of drive mechanisms used in vibrating conveyors.

9.4.1 Direct or positive motor drive with crank and connecting rod.

This is used for long heavy-duty conveyors where low frequency $(5 - 15\ Hz)$ and high amplitudes $(3 - 15\ mm)$ are required. Conveying distance (i.e. deck length) is about $2 - 50\ m$ and drive angle $25 - 35^o$ with conveying speeds of $0.2 - 0.8\ m/s$. Vibration is a problem here, but the resonance-type with counter-weight or balanced type with double or split deck will reduce vibration. They are used when compactness, light weight, high disturbing force and noiseless operation are the requirements.

9.4.2 Rotating eccentric-mass drive

Rotating eccentric-mass drive can have singe or two contra-rotating masses of equal size. The later has the advantage of synchronizing oscillation to produce linear motion perpendicular to the axes of the motors. Operating frequencies are around $15\ Hz$ $(10 - 20\ Hz)$. They do not have fixed amplitude which when determined by the load range lie between 1-10 mm. Their conveying velocities $(0.1 - 0.45\ m/s)$ and distances $(1 - 35\ m)$ are likely the same as the positive-drive conveyors. Drive angles are around $20 - 30^o$.

9.4.3 Electro-magnetic drive

Electro-magnetic drive produces simple harmonic motion at up to $50\ Hz$ using one or more electromagnets. They are amenable to step-less frequency control (using rectifier or thyristor) without interrupting operation. They provide for the directional control of vibration and operate in parallel with other vibrators. Their rubbing components are subject to rapid wear. With very small amplitudes $(0.1 - 3\ mm)$ and slow conveying velocities $(0.13 - 0.35\ m/s)$, they are most commonly used on short vibratory feeders of length $0.5 - 5.5\ m$ and drive angle $20 - 30^o$. The pull, P, produced by this drive is given by Eqn. 9.32.

$$P = mr\omega^2(\alpha^2 - 1) \qquad\qquad ...9.32$$

$where\ \alpha\ =\ \omega_o/\omega$
$=\ ratio\ of\ the\ natural\ frequency\ of\ the\ system\ actuated\ to\ the\ frequency\ of\ the\ vibrator$
$\quad r\ =\ amplitude\ of\ vibration$
$\quad m\ =\ actuated\ mass$

9.4.4 Hydraulic drive

Hydraulic drives have pneumatic or hydraulic receiver pistons fitted to the conveying deck. These pistons are driven by a remotely situated pump unit to eliminate any explosion being initiated by the electric equipment. Cone pulleys or thyristor drives are used to control speed, while manual or automatic pressure control valves on the pneumatic or hydraulic supply, control the capacity of the conveyor. This type is capable of heavy-duty work.

Table 9.6: The Average Size, Bed Depth and Conveying Speed of Some Agricultural Products

Material	Size (mm)	Bed depth (mm)	Conveying velocity (m/s)
Alumina	0.15	75	0.15
Bagasse	0.25-5.0	150	0.4
Carbon black (pellets)	1.5	75	0.18
Cement clinker	6-10	125	0.36
Cereal	6-10	150	0.36
Coal	18-26	125	0.3
Crumb rubber	6	100	0.3
Detergent powder	0.15	75	0.25
Gravel	6-10	125	0.33
Limestone	10-30	100	0.36
Milk powder	0.075	35	0.13
Plastic pellets	3-6	100	0.36
Sand-damp	0.8	100	0.4-0.45
Sand-dry	0.8	75	0.25-0.3
Salt (table)	0.4-0.8	50	0.3
Sugar (granulated)	0.5-0.8	60	0.25
Tobacco (cut)	4-8	250	0.36
Wood Chips	10	250	0.4

9.5   POWER REQUIREMENT

Each of the above drive mechanisms will require different amplitudes, frequency ratios ω_o/ω and exciter forces and will result in different power requirements. To enable the vibratory conveyors to perform their intended function of feeding, transporting, screening, grinding, polishing, de-burring, classifying, etc., the energy must be transmitted in the form of vibrations. If all the design parameters of the different drive types of the vibratory conveyor are taken into consideration, the precise computation of power requirement is very complex (Colijn and Rademacher, 1978; Alexandrov, 1981; Colijn 1985). However, based on average factors of power consumption in conveying materials over a distance of 1 m, factors K_3 and K_4 (depending on conveyor length) are used in the following equations (Eqns. 9.33 and 9.34) for calculating power ratings of the drive motor in kW.

$$For \ L \leq 10m, HP = (CQ/\eta)[10K_3L + (H/367)] \qquad ...9.33$$

$$For \ L > 10m, HP = (CQ/\eta)[10K_3 + (L-10)K_4 + (H/367)] \quad ...9.34$$

where Q = capacity in $\frac{t}{h}$; H = vertical lift of load in m;

($H = 0$ for horizontal conveyors and downgrades);

η = efficiency of drive $(0.95 - 0.97)$.

C = conveyability factor $\left(\begin{array}{l}1 \text{ for granular and lumpy materials;} \\ \text{and } 1.5 - 2 \text{ for pulverized and dusty loads}\end{array}\right)$;

K_3 and K_4 = length factors (Table 9.7).

Table 9.7: **K-values (Factors of Specific Power Consumption)**

Type of Vibrating Conveyor	Calculated Capacity t/h	K_3 ($\times 10^{-3}$)	K_4
Suspended from shock absorbers, actuated by unbalanced pulley.	5 - 50	6-7	-
	over 50	5-5.5	-
Unbalanced, supported by inclined vibrating arms with stiff connecting rods.	5 - 50	7-10	5-6
	over 50	5-6	3.5-4.5
Balanced, single or double-trough, actuated by eccentric shaft with elastic connecting rods.	5 - 50	10 - 12	8-10
	5 - 50	4.5 - 5	3.5-4
	over 50	4 - 5	3-3.5

9.6 APPLICATIONS OF VIBRATORY CONVEYORS

Vibratory conveyors have unique capabilities. They are versatile, of relatively low maintenance and ease of installation. Because of the fact that the conveyed material moves independent of the conveying medium in this conveyor, unlike in the other conventional conveyors, the vibratory conveyor is applied in solving many difficult material handling problems. It is suitable for handling a very wide range of material types from granular to pulverized in horizontal (or near horizontal) movement up to 15° inclines and with some kind of processing operation (cooling, mixing, segregation, de-watering, etc.) taking place as the material is being transported. Hill (1980) enumerated some important applications of the vibrating conveyor as follows:

- Different materials can be blended after feeding them separately to the conveyor.
- Scalping and screening can be performed while conveying.
- Manual picking of contaminants can be done while conveying because of slow speed and continuous tumbling of materials in the conveyor.
- Cooling, heating and even washing of products can be done by having a deck with perforated floor or side walls.

- Different products can be handled on a single unit by use of divided decks and material can be distributed through a number of discharge points.
- Extremely hot and abrasive materials can be handled by the vibrating conveyor.

Other important attributes of the vibrating conveyor include:
- Self-cleaning and leak proof without slippage
- Good sanitation can easily be achieved by making the deck dust-tight.
- No moving parts, except the deck, in contact with the material.

9.7   SPIRAL CONVEYORS/ELEVATORS

This upright vibrating conveyor has the conveying deck wound helically, at shallow angles, on a vibrating central support core. It lifts the material up to about 10 m by the particles hopping along tangents. With diameter varying between 120 – 1000 mm, its maximum capacity is up to 20 t/h. They have the advantage that some processing applications that involve heat transfer are readily adaptable to them because of long contact surface and long transit time. They also occupy very little floor space, being compact. Air- or gas-tightness is achievable using the spiral conveyor (Haneman and Mocha, 1978) for track length of up to 26 m.

9.8   SUMMARY

1. Vibratory conveyors can move lumpy, granular and some fine materials on upgrades of between 5 - 20°. Dust recovered by means of cyclones and materials of the same size cannot be handled.
2. Conveying speeds range 0.1 - 0.5 m/s depending on the properties of the material such as:- bulk weight, size and shape of particles, moisture content (10% moisture content facilitates conveying while higher values reduce speed), elastic constants, cohesion and angle of internal friction, permeability by air, etc.
3. Achievable amplitudes depend on the exciters: 0.1 - 2.0 mm for electromagnetic exciters; 0.5-5.0 mm for unbalanced pulleys; and 3-12 mm for eccentric shafts. The maximum length of balanced vibratory conveyors is 100 m.
4. Compared to other conveyor types, the vibratory conveyors are lighter and more economical. However, they cannot compete with belt conveyors in points of weight and power consumption for distances over 50 m.
5. Oscillating conveyors, usually mounted on rearward-inclined cantilever supports, move materials in a hopping or bouncing movement at about 0.25 m/s with less wear on the trough or deck.
6. Oscillating conveyors are driven by eccentric motors at 400 – 500 rpm with unit length of about 30m and capacities up to 90 t/h.
7. Four types of motion occur in oscillating conveyors depending on deck frequency. Most agricultural operations use the moderate frequency (type III) with pure slip motion where downward vertical motion equals gravitational acceleration.

8. Crank-operated variable pressure oscillating conveyors used in agricultural engineering practice have r = 30 – 40 mm; average travel speed of material = 0.15 - 2.0 m/s; depth of material in the trough ≤ 100 mm; maximum incline < 15° and efficiency = 0.80 - 0.98.

9. Drag-link operated constant pressure oscillating conveyors have capacities up to 80 t/h, with material depth = 50 – 100 mm; maximum velocity = 0.2 m/s; N = 50 – 100 rpm; r = 50 – 150 mm and capacity factor = 0.5 - 0.9 depending on shape of deck.

9.9 SOLVED PROBLEMS

Q.9.1 An open oscillating conveyor of cross-sectional area 0.01 m² transports material of bulk density 0.74 t/m³ at a speed of 0.15 m/s over a distance of 4.8 m. If the oscillating arm is 40 mm, tilted at 20°, has frequency 300 cycles/min., find the friction coefficient required to design a 70-kW conveyor motor with efficiency of 90%. What will be the optimum deck acceleration for the material to continue in motion during a forward stroke? During a return stroke, will the permanent load-to-deck contact be maintained by the conveyor?

Data: $\varrho = 0.74\ t/m^3, A = 0.01\ m^2,\quad V = 0.15\ m/s,\quad L = 4.8\ m, r = 40\ mm;$
$N = 300\ cycles/min;\ \alpha = 20°;\ HP = 70\ kW,\ \eta = 90\%$

Solution:

(a)   Using Eqn. 9.12,

$$HP = (4x1.0^{-4}QL/\eta\ Tan\ \alpha)[(6x1.0^{-4}rN^2)/f_s + (1)]\ kW$$

$$Q = 3600A\psi V\varrho_{loose}\ (Eqn.\ 9.19)$$

$$Choose\ \psi\ = 0.6\ for\ open\ decks\ and\ \ \varrho_{loose} = 0.8\varrho$$

$$\therefore Q = 3600x0.01x0.6x0.15x0.8x0.74 = \mathbf{1.918\ t/h}$$

Substitute in Eqn. 9.12 to find f_s

$$70 = (4x1.0^{-4}x1.918x4.8/0.9xTan\ 20°)[(6x1.0^{-4}x40x\ 300^2/f_s) + 1]$$

$$70 = 0.01124[(2160/f_s) + 1]$$

$$\therefore f_s = \mathbf{0.347}$$

(b)   Condition for load to continue to slide forward during a forward stroke is Eqn. 9.7

$$j_{deck} > f_s g/Cos\ \alpha - f_s Sin\ \alpha = 0.347x9.81/0.9397 - (0.347x0.342) = \mathbf{4.146\ m/s^2}$$

$$j_{deck}(optimum) = \mathbf{4.15\ m/s^2}$$

(c)For permanent load-to-deck contact (use Eqn. 9.10): $j_{deck} < g$

$$j_{deck\ y} = j_{deck}Sin\ \alpha = 4.15Sin\ 20° = \mathbf{1.42m/s^2}$$

Since 1.42m/s² < 9.81m/s², permanent load-to-deck contact is maintained.

Q.9.2   For a variable pressure oscillating conveyor, the amplitude is 25 mm, the drive angle = 15° and the load continues in motion during forward stroke,

i.   What will be the least rotational speed of the crankshaft if f_s = 0.28?
ii.   At 350 rpm, will there be load-to deck contact?
iii.   What rotational speed will the motor have to develop 42.5kW for conveying 1.8t/h grain over 6m? (Assume η = 0.82)

Data: $r = 25\,mm$; $\alpha = 15°$; $f_s = 0.28$; $L = 6m$; $Q = 1.8\,t/h$; $\eta = 0.82$; $HP = 42.5\,kW$

Solution:
(a)   The least rotational speed is taken from Eqn. 9.7

$$j_{deck} = N^2 r/90 > f_s g/Cos\,\alpha - f_s Sin\,\alpha$$

$$\therefore N > \sqrt{90 f_s g/r[Cos\,\alpha - f_s Sin\,\alpha]} = \sqrt{90 x 0.28 x 9.81/0.025[0.966 - (0.28 x 0.259)]}$$

$$= 105.2 rpm$$

(b)   At $350 rpm$, $j_{deck} = 350^2 x 0.025/90 = 34.03 m/s^2$

And $j_{deck\,y} = j_{deck}\,Sin\,\alpha = 34.03 Sin\,15° = 8.81 m/s^2$

Using Eqn. 9.11, **since $j_{deck\,y}$ = 8.81m/s² < g = 9.81m/s², there is load-to-deck contact.**

(c)   The power required for a variable pressure oscillating conveyor is given by Eqn. 9.12

$$HP = (4 x 1.0^{-4} QL/\eta\,Tan\,\alpha)[(6 x 1.0^{-4} rN^2/f_s) + 1]\ \ kW$$

$$42.5 = (4 x 1.0^{-4} x 1.8 x 6/0.82\,Tan\,15°)[(6 x 1.0^{-4} x 25 N^2/0.28) + 1]$$

$$42.5 = 0.0197(0.0536 N^2 + 1)$$

The rotational speed that will develop the required power is got by solving the last equation above.

$$N = 200.6 rpm$$

Q.9.3 A 40 t/h vibrating conveyor has length L = 15.6 m. What is the power of the 95% efficient drive motor under the following conditions? (i) Horizontal movement; (ii) Moving 15° upgrade and (iii) Moving 10^odowngrade. Assume conveyability factor = 1.0; power coefficients $K_3 = 8.4 \times 1.0^{-3}$ and $K_4 = 5.5 \times 1.0^{-3}$

$Data: Q = 40\ t/h;\ L = 15.6\ m;\ \eta = 95\%;\ \beta = 0^o, 15^o, 10^o;\ C = 1;$
$\quad K_3 = 8.4\ x\ 1.0^{-3}\ and\ K_4 = 5.5\ x\ 1.0^{-3}$

Solution:
Power equation for vibrating conveyor is given by Eqn. 9.34

$$HP = CQ/\eta\ [10K_3 + (L - 10)K_4 + (H/367)]$$

(a) For horizontal movement; $L = 15.6\ m;\ H = 0;\ \beta = 0^o$

$\therefore HP = 1x40/0.95\ [10x8.4x1.0^{-3} + 5.6x5.5x1.0^{-3}] = \textbf{4.83}\textbf{kW}$

(b) $For\ \beta = 15^o\ upgrade: L = 15.6\ Cos\ 15^o = 15.07\ m, and$
$\quad H = 15.6\ Sin\ 15^o = \textbf{4.04}\ \textbf{m}$

$\therefore HP = 1x40/0.95\ [10x8.41.0^{-3} + 5.07x5.5x1.0^{-3} + (4.04/367)] = \textbf{5.175}\textbf{kW}$

(d) $For\ \beta = 10^o\ downgrade, L = 15.6\ Cos\ 10^o = \textbf{15.36}\ \textbf{m}; H = 0$

$\therefore HP = 1x40/0.95\ [10x8.4x1.0^{-3} + 5.36x5.5x1.0^{-3}] = \textbf{4.778}\textbf{kW}$

Q.9.4 The flight ratio of a vibratory feeder conveyor is 1, find throw factor and the machine number if the drive angle is 50°. Calculate V_{max}, V_{ave} and V_{actual} and Q if the velocity efficiency is 60% when conveying dry grain of depth 120 mm and width 350 mm and bulk density 1.25 t/m^3 with amplitude = 15 mm.

$Data: n = 1, \beta = 50^o;\ \eta_v = 0.60;\ \varrho = 1.25\ t/m^3;\ h = 120\ mm;\ r = 15\ mm,$
$\quad \beta = 350\ mm;\ A = 0.042\ m^2$

Solution:
(a) From Eqn. 9.22; K_v = throw factor =

$$K_v = \sqrt{1 + ((Cos\ 2\pi n + 2(\pi n)^2 - 1)/(2\pi n - Sin(2\pi n)))^2}$$

$$= \sqrt{1 + ((Cos\ 2\pi + 2(\pi)^2 - 1)/(2\pi - Sin\ 2\pi))^2}\quad = \textbf{3.297}$$

(b)   From Eqn. 9.25, the machine number K = K_v/Sin ß

$\therefore K = 3.297/Sin\ 50^o = \mathbf{4.304}$

This shows that the feeder conveyor can withstand high frequency vibrations.

(c)   From Eqn. 9.21,

$$\omega = \sqrt{K_v g/rSin\ \beta} = \sqrt{3.297 x 9.81/15 x 1.0^{-3} Sin\ 50} \qquad \mathbf{= 53.05\ Hz}$$

$$\therefore N = \omega/2\pi = 53.05/2\pi = \mathbf{8.444 rps = 506.6\ rpm}$$

From Eqn. 9.27; $V_{max} = \omega r Cos\ \beta = 53.05 x 15 x 1.0^{-3} Cos\ 50^o = \mathbf{0.51\ m/s}$

From Eqn. 9.28; $V_{ave} = \eta_v V_{max} = 0.6 x 0.51 = \mathbf{0.31\ m/s}$

Choosing $f_m = 0.85$; $f_h = 1.0$ and $f_s = 1.0$, we substitute in Eqn. 9.29

$$V_{actual} = \eta_v f_m f_h f_s r (2\pi N) Cos\ \beta$$

$$V_{actual} = 0.6 x 0.85 x 1.0 x 1.0 x 15 x 10^{-3} x 2 x \pi x 8.444 Cos\ 50^o = \mathbf{0.26\ m/s}$$

From Eqn. 9.31; $Q = \varrho A V_{actual} = 1.25 x 0.042 x 0.26 x 3600 = \mathbf{49.14\ t/h}$

Q.9.5 A single deck vibrating conveyor has centrifugal drive of amplitude 1.24mm, and conveys dry granular load , Q = 45 t/h at drive angle β = 30°. Operating at 1500 cycles/min, find speed of load if the conveyor is moving (i) horizontal, (ii) 15° upgrade, and (iii) 10° downgrade

Data: $r = 1.24\ mm$; $N = 1500\ rpm$; $\beta = 30^o$, $Q = 45\ t/h$;
    $\alpha = 0$; $15^o\ up$; $10^o\ down$; $\omega = 2\pi N = 157.08\ Hz$.

Solution:
From Table 9.2, choose $K_v = 2.8$. From Table 9.4 choose $K_1 = 0.95$ and $K_2 = 2.25$

From Eqn. 9.17; $V = (K_1 + K_2 Sin\ \alpha) r \omega Cos\ \beta \sqrt{(1 - (1/K_v^2))}$

(a)   For horizontal, $\alpha = 0$

$$\therefore V = (0.95)x1.24x10^{-3}x157.08Cos\ 30^{\circ}\sqrt{\left(1 - (1/2.8^2)\right)}$$

$$\therefore V = 0.1602\sqrt{(1 - 0.128)} = \mathbf{0.15\ m/s}$$

(b)   For 15° upgrade, α = 15°,

$$V = (0.95 - 2.25Sin\ 15^{\circ})x1.24x10^{-3}x157.08Cos\ 30^{\circ}\sqrt{\left(1 - (1/2.8^2)\right)}$$

$$\therefore V = 0.3677x0.1687\sqrt{(1 - 0.128)} = \mathbf{0.058\ m/s}$$

(c)   For 10° downgrade, α = 10°

$$V = (0.95 + 2.25Sin\ 10^{\circ})x1.24x10^{-3}x157.08Cos\ 30^{\circ}\sqrt{\left(1 - (1/2.8^2)\right)}$$

$$\therefore V = 1.3407x0.1687\sqrt{(1 - 0.128)} = \mathbf{0.211\ m/s}$$

CHAPTER 9 Review Questions

9.1 What is shaking conveyors, what do they consist of and where are they used in an agricultural engineering enterprise?

9.2 What are the advantages and disadvantages of shaking conveyors?

9.3 Describe, with the aid of a diagram, an oscillating conveyor. How does it function? How is it completely specified?

9.4 Using deck frequency, give the four different kinds of motion of an oscillating conveyor. Which of these is encountered in agricultural operations?

9.5 Derive the equation of deck acceleration in both forward and return motion of a variable pressure oscillating conveyor.

9.6 How does the load on the deck of a constant pressure oscillating conveyor move through one cycle?

9.7 What three problems are interrelated in oscillating conveyor calculations?

9.8 What are the differences between vibrating and oscillating conveyors? What are the advantages and disadvantages of both conveyors?

9.9 What deck- and material-related factors influence the materials' movement on a vibratory conveyor?

9.10 Define the following terms: throw factor, machine number, velocity - efficiency; and give their corresponding equations.

9.11 State some factors influencing design parameters of shaking conveyors.

9.12 What makes the vibrating conveyors unique? How are they employed in materials handling problems?

9.13 Give the expressions for the load speed, throughput capacity and power of oscillating conveyors.

9.14 What is the operating cycle of an intermittent motion conveyor?

9.15 An oscillating conveyor has $\omega = 45\,Hz$; $\beta = 20^o$; $\alpha = 25^o$; $r = 35\,mm, K_1 = 0.85$; how does the load move and at what speed? If the load has density $= 0.65\,t/m^3$, $A = 0.40\,m^2$ and $\psi = 0.6$, find the throughput capacity *($K_v = 2.73$ [Load moves with jumps] $V = 0.12\ m/s$; $Q = 55.2\ t/h$)*

9.16 *A 1.5 Hz oscillating conveyor of $r = 25\,mm, \alpha = 35^o$ carried $8\ m^3/h$ maize of $= 1.15\ t/m^3$ across a drier of $L = 10\ m$ at $V = 0.225\ m/s$. If $f = 0.6$ and $\eta = 0.90$, find the design HP. If however, f varies from 0.25 to 0.65 but HP remains constant, draw a graph of f vs N. For what f is $N = 77$rpm? (HP = 11.88 kW; f = 0.44).*

9.17 For an oscillating conveyor, $L = 5.4\ m, N = 400\ rpm, V = 0.20\ m/s$; $r = 30\ mm, f = 0.4, \alpha = 20^o$ and transports $1.2\ t/h$ of grain. What is the power if $\eta = 80\%$? If $j_{deck} = 8 - 10V + 3V^2$ will the particle continue its motion during forward and return strokes respectively? Will there be permanent load to deck contact? *(HP = 50.03 kW; j_{deck} = 6.12 m/s²; $j_{deck.f}$ = 5.32 m/s²; $j_{deck.r}$ = 3.65 m/s², motion is continued during both strokes; $j_{deck.y}$ = 2.59 m/s² permanent load-to-deck contact exists)*

9.18 For an oscillating conveyor, $A = 0.006\ m^2, V = 0.15\ m/s.$; $L = 5\ m$; $r = 30\ mm, N = 200\ rpm, \alpha = 20^o$. If the motor develops 70 kW at 85% efficiency, find

f. Show graphically how the motor power will vary for different crops of f = 0.20 to 0.45 at 0.05 increments. (*f = 0.27; parabolic graph*).

9.19 Given that for a vibrating conveyor, $\omega_o = 3000\ rpm, \omega = 2400\ rpm, m = 10\ kg\ and\ a = 1.5\ mm$, what will be the pull produced by the vibrating motor?
(*P = 13.5 N*)

9.20 A vibrating conveyor has the following parameters, $a = 3\ mm$; $\omega = 120\ m^{-1}$; $\beta = 35^o$; $\alpha = 8^o$; $K_1 = 0.8$; $K_2 = 1.8$; $K_3 = 5.5\ and\ K_4 = 3.0$; $L = 50\ m$. If Q = 60 t/h of grain is moved, what is the load speed and drive motor power if $\eta = 95\%$?
(*C = 2.55; V = 0.149 m/s; HP = 12.25 kW*).

10

██████████████████████████████████████

PNEUMATIC AND HYDRAULIC CONVEYORS

10.1 INTRODUCTION

These conveyors deal with the movement of bulk materials (powder, short fibre and granules) using air or water as the carrier over pipelines and under pressure. The main purpose of a pneumatic conveyor is to move materials from, may be, a bulk transport vehicle to a storage hopper, or from a storage hopper to a bagging machine using air (or gas) pressure. It transports dry, free-flowing, granular materials in suspension within a pipe or duct by means of a high-velocity airstream or by the energy of expanding compressed air. They convey materials over more than 2 km at a rate of 400 t/h and to a height of 100 m. They are used for (a) dust collection (b) conveying soft materials, such as grain, dry foodstuffs (flour and feeds), chemicals, woodchips, carbon black, and sawdust, (c) conveying hard materials (hay and straw).

Hydraulic conveying involves moving solid particles in suspension in a moving liquid so long as the materials do not dissolve in the liquid or is affected adversely by the liquid. Hydraulic handling is an economic means of transport in terms of injury to fresh fruits and vegetables and also in the process of washing.

When considering either pneumatic or hydraulic conveyors, one has to have the basic principles of fluid mechanics at the back of one's mind. This would necessitate the knowledge of the aerodynamic and hydrodynamic characteristics of the agricultural products, i.e. the behaviour of the products when they are in fluid medium.

10.2 AERODYNAMIC PROPERTIES OF THE FLUID/SOLIDS SYSTEMS

The behaviour of particles moving in or being moved by a fluid is characterized by their aerodynamic parameters of critical (terminal) velocity, coefficient of resistance of the fluid and the drag coefficient (Klenin et al., 1986). In a vertical fluid flow (Fig. 10.1), three forces are acting on the particle falling under gravity in the fluid: the gravitational force, F_g.; the buoyancy force, F_b; and the drag force, F_D. The force of gravity on a particle is calculated using the mass of the particle. The buoyancy force can be calculated using the particle density and the density of the fluid. The drag force is based upon its velocity, the density of the surrounding fluid, and the Reynolds number of the flow around the particle.

For equilibrium and steady state motion of the particle in the fluid, the sum of the upward forces (F_R) equals the gravity force (F_g) and is given by Eqn. 10.1.

$$F_R = F_D + F_b = F_g \qquad \text{...10.1}$$

where F_R = sum of upward forces ($F_D + F_b$) which from Newton's equation gives Eqn. 10.2.

$$F_R = CA\varrho_f(V_f - U)^2 \quad N \qquad \text{... 10.2}$$

where

ϱ_f = density of fluid (kg/m^3);

A = projected area of the body on a plane perpendicular to the direction of motion of the fluid stream (m^2),

V_f = velocity of fluid stream (m/sec.),

U = velocity of the particles (m/s), and

C = coefficient of resistance.

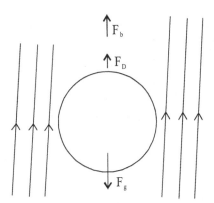

Fig. 10.1: Forces on a particle falling through a stream of fluid

Since F_R and F_g act in opposite directions, the particle may move downward when $F_g > F_R$, upwards when $F_R > F_g$ or may remain suspended when $F_R = F_g$ and U = O. The terminal velocity, V_t, is that constant fluid velocity a body falling inside a fluid will attain when the total sum of upward forces (F_R) equals the gravitational force (F_g). This velocity is important in such cases as winnowing, and must be exceeded for the particle to be transported by the fluid. For spherical particles, the drag force is given as in Eqn. 10.3.

$$F_D = F_g - F_b = \frac{\pi}{6}d^3 g(\rho_p - \rho_f) \qquad \text{...10.3}$$

where

d = diameter of particle (m); g = acceleration due to gravity m/s^2;

ϱ_p = density of particle (kg/m^3); ϱ_f = density of fluid (kg/m^3).

Using dimensional analysis, the drag force (Eqn. 10.4) can be written as: -

$$F_D = C_D A \varrho_f V_t^2/2 = C_D \pi d^2 \varrho_f v_t^2/4x2 \qquad \qquad \text{...10.4}$$

where

$C_D = drag\ coefficient;\ d = particle\ diameter\ (m);$
$A = area\ of\ particle\ (m^2);\ V_t = terminal\ velocity\ (m/s).$

Equation 10.4 is similar to Eqn. 10.2 with U = 0. Thus, the terminal velocity of spherical particles got by equating Eqn.10.3 and Eqn.10.4 is given by Eqn. 10.5.

$$V_t = [4gd(\varrho_p - \varrho_f)/3\varrho_f C_D]^{\frac{1}{2}} \qquad \qquad \text{...10.5}$$

However, the drag coefficient, C_D must be known in order to use Eqn.10.5. It is given in Eqn. 10.6.

$$C_D = 4gd(\varrho_p - \varrho_f)/3\varrho_f V_t^2 \qquad \qquad \text{...10.6}$$

showing C_D as a function of V_t^2 It is difficult to determine any of them unless by trial and error. Using dimensional analysis and Stokes's Law, it can be shown that C_D is a function of Reynolds number, R_e, which is a ratio of inertia force to the viscous force on the particle, and the shape of the particle around which flow occurs. For spherical particles, it is given as Eqns. 10.7 and 10.8.

$$C_D = 24/R_e \qquad \qquad \text{...10.7}$$

$$and \quad R_e = V_t d/v \qquad \qquad \text{...10.8}$$

where

$v =$ kinetic viscosity.

When R_e is plotted against $C_D R_e{}^2$, the terminal velocity can be determined since from Eqn. 10.6 and Eqn. 10.8 as Eqn. 10.9.

$$C_D R_e^2 = 4gd^3(\varrho_p - \varrho_f)/3\varrho_f v^2 \qquad \qquad \text{...10.9}$$

which is independent of velocity.

For irregularly shaped agricultural products which assume their own random configurations in the moving fluid, the analysis given above could be applied when one determines an "equivalent diameter" or the Stokes diameter of the product. This is the diameter of a spherical particle of the same density as the product and would have the same terminal velocity in the liquid. Hawksley (1951) suggested the introduction of sphericity ϕ_s to Eqn.10.6 and Eqn.10.8 to obtain the apparent drag coefficient (C_D^1) (Eqn. 10.10) and Reynolds number (R_e^1) (Eqn. 10.11) for irregularly shaped particles.

$$C_D^1 = \phi_s\, 4gd_e(\varrho_p - \varrho_f)/3\varrho_f V_t^2 \qquad \text{...10.10}$$

$$R_e^1 = \left(1/\sqrt{\phi_s}\right).(V_t d_e/\nu) \qquad \text{...10.11}$$

They are called apparent because they are measured for a given set of geometrical and hydraulic conditions. Both coefficient of resistance (C) and drag coefficient (C_D) are functions of the surface characteristics, the size and shape of the product, the state and species of the medium in which it is located, and on the velocity of the medium. The values of these quantities for some agricultural products are given in Table 10.1 (Keck and Goss, 1965; Klenin et al., 1986; ASAE Data D241.2, 2005).

Table 10.1   Aerodynamic Properties of Some Crops

Crop	Critical Velocity V_t (m/sec)	Drag Coefficient C_D	Coefficient of Resistance C	Bulk Density (kg/m³)	Particle Density (kg/m³)	Equivalent Particle Diameter (mm)
Wheat	8. 9-11.5	0.075-0.52	0.184-0.265	769	1300	4.08
Corn	11.48-14.0	0. 05-0.70	0.16 -0.28	718	1390	7.26
Oats	6.68-9.11	0.169-0.51	0.118-0.15	410	1050	4.19
Rye	8.36-9.89	0.1 - 0.14	0.16 -0.22			
Soybean	11.5-14.5	0.20-0.50		769	1180	6.74
Barley	5.4 - 8.3	0.22 - 0.61		615	1130	4.05
Wheat Straw						
0.6cm	5.15	0.84				
2.5cm	4.25	0.8				
7.5cm	3	0.9				
25.0cm	2.7	0.91				
Potatoes	32	0.64				

10.3   PNEUMATIC CONVEYOR

This is the transportation of dry bulk particulate or granular materials (powder, short fiber, granules) through a pipeline by a stream of gas or air at high-velocity.   Pneumatic transport is frequently used in agriculture to convey granular materials (Sitkei, 1986). It can convey dusts, fibers, sand, rags, cotton etc. (Henderson and Perry, 1955). Pneumatic conveying systems transport powders, granules, dry bulk, and flakes through an enclosed conveying pipeline using pressure differential and gas flow (usually air) generated by an air movement device, for example, a roots blower, fan, or compressor. Pneumatic conveyors provide a cost-effective manner to handle and transfer powdered and bulk granular materials quickly with very little loss. They are suitable for a range of process industries such as; food and beverage, pet food, chemicals and detergents, renewables, and specialist materials.

These conveying systems require only a source of a blower or compressed air (air moving system), a means of feeding the material into the pipeline (feeding system), and a receiving hopper (discharge system), for separating the conveyed material from the conveying air. Pipelines and fittings connect these systems to make the conveyor. They are used extensively in many industries: pharmaceutical, food, chemical, glass, and cement, plastics, mining and metal including provision for storage, transport, recovery and mastering of the materials. Materials such as asbestos powder, plastics, coffee beans, sawdust, molding sand, grass seed, grain, talcum powder, flour and sugar, whole fish, and live chicken etc. can be conveyed using pneumatic system (Woodcock and Mason, 1987; Dixon, 1981). Pneumatic conveyors are more flexible than augers (Roth and Field, 1991) and bucket conveyers (Rothwell et al., 1991) because the duct does not need to be in a straight line. The power requirement for pneumatic conveyors is much greater than for a mechanical conveyor of equal capacity but the duct can be led along any path. There are no moving parts and no risk of injury to the attendant. The performance cannot be predicted with the accuracy usual with the various types of mechanical conveyors and elevators (Altamur and Hawkins, 1986). Also, the pneumatic conveyor is used where a straight run cannot be obtained or where the inlet has to be moved very often, as in emptying a large flat-bottomed silo.

However, the important thing is that the conveying air velocity will be much higher than the terminal velocity of the material thereby carrying the material in suspension and causing no corrosion or erosion on pipe walls except at bends. Also, the pressure differential in the pipeline must be sufficiently high to accelerate the mixture of air and material to the required velocity. Generated pressure is as high as 8 kg/cm^2. High pressure conveying system has been used in fully automatic feed preparation and distribution system (Puckett, 1960).

Advantages
The following are some of the advantages of pneumatic conveyors:

–   The delivery of materials is over a path capable of changing its direction in any plane as required (flexibility) i.e. adaptable to irregular layouts.
–   The conveyed material can be processed simultaneously (washing, heating, cooling, drying, etc.) with its conveying.

- A single pneumatic conveying system can serve an almost limitless number of loading and unloading points.
- The conveyor system is self-cleaning with minimum of human attendance.
- There are improved labour conditions with no dust nuisances or dust hazards because of air- and gas-tightness.
- In the system, conveying is almost totally automated with mechanical simplicity (only one moving part - the fan, eliminating danger to the worker), and low initial and maintenance costs.
- There is considerable reduction of losses of material via spillage and wastage.
- There are wide varieties of materials that can be conveyed (dusts, fibers, sand, grain, rags, cotton, etc.) using the pneumatic conveyor.

Disadvantages

The pneumatic conveyor has the following disadvantages:
- It has high power requirement (1 - 5 kWh/t i.e. 10-15 times higher than mechanically-driven ones (CFIA, 1978).
- There is low output i.e. low throughput capacity.
- There is rapid wear of equipment by the conveyed material.
- It is noisier than augers.
- There is the problem of dust recovery from exhaust air before this leaves into the atmosphere.
- There is the inability of pneumatic conveyors to convey wet, caking and sticky loads.
- There is possible damage to conveyed material during transport.
- There is explosion risk with certain products.

10.3.1 Modes of Flow in Pneumatic Pipelines

Particles are moved in pneumatic pipelines in two major ways: -

(a) Dilute-Phase or Lean-Phase Flow

Here, conveying air or gas velocities are relatively high, with the particles carried in uniform suspension at low concentrations (Srivastava et al., 1996). The solids loading ratios (i.e. the ratio of a material mass flow rate to a fluid mass flow rate) are less than 15. Particles are fully and uniformly suspended and are widely spaced at about eight diameters apart. Flow is ensured when the conveying velocity does not fall below 13-15 m/s for most bulk materials. Air velocity is generally 20 – 25 m/s with low static pressure of usually 50 mbar upwards. Conveying of agricultural materials is done in this phase.

(b) Dense-phase Flow

When conveying velocities are less than terminal velocities required to keep the material in suspension, particles begin to settle at the bottom of the pipeline and are conveyed as discrete masses or clusters forming the dense-phase flow. The solids loading ratios are greater than 25 reaching several hundred for some materials (Wen, 1971). With high load ratios, the conveying gas percolates through the bulk material in the pipe yielding high pressure loss and tears off pipe material at bends. The main problem is that reduction in conveying velocity is minimal otherwise blockages of pipelines will occur mostly at bends. Air velocity is about 0.5-2.0 m/s requiring high static pressure.

10.3.2 Types of Pneumatic Conveying Systems

There are three main types of pneumatic conveying systems: - low-pressure systems, high-pressure systems, and low-velocity systems.

(a)   **Low-Pressure Systems**

They convey materials at relatively very low pressures using high-velocity low-density air with conveying velocity ranging from 10.2 m/s for saw-dust to 20.3 m/s for grain. For some materials a velocity of 30 m/s may be required to keep the materials suspended. They can handle a wide range of pulverized, granular and fibrous materials being very versatile, economical and flexible. There are three types of this system: the positive-pressure, the negative (vacuum), and the combination types.

The positive-pressure type (Fig. 10.2) uses a centrifugal fan or blower of pressure (low to moderate) about 1 bar, to blow air/gas into the pipeline while the material is fed at one feed point by means of an air-lock ahead of the fan/blower and conveyed along the pipeline to a number of discharging units each containing a cyclone/filter separator and a receiving hopper. The fan creates a pressure head which carries the suspended material to a terminal separator. The problems with the system include: -

(i)   Leakage of air gas from valves.
(ii)   Irregular feeding.
(iii)   Clogged filter or blocked vent line.
(iv)   System is limited to 70 kPa pressure.

Since abrasion is no problem, steel or galvanized-metal pipe ducts are satisfactory. However, unnecessary bends and fittings must be avoided in order to minimize power consumption (Segler, 1951).

The negative-pressure (vacuum) systems (Fig. 10.3) convey materials on the suction (pressure about 0.7 bar or 0.4 - 0.5 Atm.) with the suction fan positioned at the discharge end of the pipeline. It is well suited for handling ingredients fed from several hoppers into a single process line or collection point. It is important to have a highly efficient discharging unit so as to minimize damage to the air mover. Also, since air leakage is inwards, the injection of dust into the surrounding is eliminated but inflowing air could result in unwanted contamination of conveyed product and reduction of air at inlet ends of pipeline. So, leakages must be minimized. However, the negative pressure type is not common in agricultural engineering applications or operations.

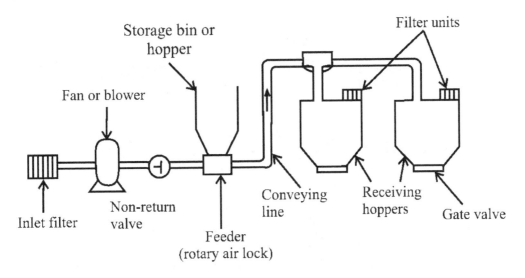

Fig. 10.2: Positive - pressure pneumatic system

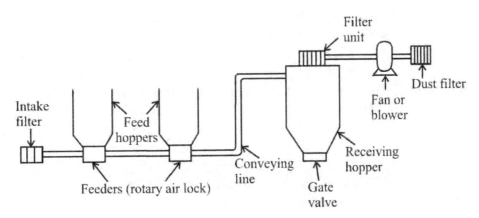

Fig. 10.3: Negative - pressure (vacuum) pneumatic system

In the combination system (suck-blow or pull-push) (Fig. 10.4) materials are picked up by suction from a number of loading points, passed through a rotary air lock into the pressure system which is used to deliver the materials to a number of unloading points over a long distance. The air mover for this combination system is larger than either of the above systems for obvious reasons - doing the work of suction and conveying at the same time. It is essential: -

(i)   To design the system so that available power is shared adequately between the suction side and the positive pressure side on account of different operating pressures and pressure losses,

(ii)   To protect the fan or blower against ingress of solid particles.

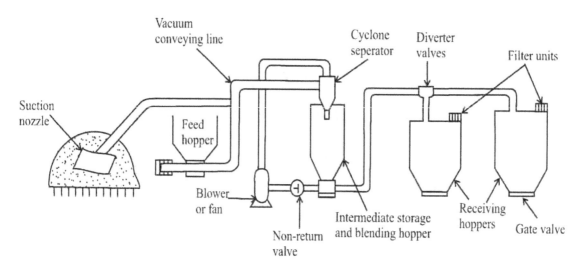

Fig. 10.4: Combined positive - and negative - pressure system

(b)   **High-Pressure and Low-Velocity Systems**

These two systems are used mainly for the dense-flow phase in order to overcome some of the limitations of the low-pressure system which include high power consumption and high solids' velocity. While these systems save running costs by lower air consumption and use of smaller filters, they increase capital cost by the introduction of compressors (or positive displacement blowers) that would generate the required high pressures (7 – 8 bars) and the blow tank. They are used to convey materials over distances greater than 2.5 km (Marcus and Rizk, 1985) at velocities less than 1 m/s (Mainwaring and Reed, 1986). They are not, however, common in agricultural processing operations.

10.3.3 Components of Pneumatic Conveying Systems

There are four main components in pneumatic conveying systems: -

(a)   **Air Supply**

The air mover is the most important component of the pneumatic conveying system. The air movers range from fans and blowers to compressors and vacuum pumps which produce high volumetric flow rates which depend upon the conveying velocity and the size of the pipeline. The selection of the air mover depends upon the pressure and air flow requirements of the system. For pressure up to 68.9 kg/cm^2 a centrifugal blower is more tolerant of dirt.

The fan is meant to overcome the following resistance:
(i)   Suctions of the various hoods,
(ii)   Collector loss and
(iii)   Loss due to pipe friction.

The pipe loss comprises the sum of losses in corresponding branches and in the main pipe. And the total loss in the system is the sum of the three resistances mentioned above.

(b) **Feeding Devices** - are capable of introducing the product reliably at a constant rate between the conveying line and the supply hopper. They include rotary valves, screw feeders, venture feeders, gate lock valves, suction nozzle and blow tanks. The design of these devices depends upon the type of conveying system used. For rotary feeders, the product mass flow rate M_s is given by Eqn. 10.12.

$$M_s = \varrho_b \, V n \, N \qquad\qquad\qquad ... 10.12$$

where

$\varrho_b = bulk \; density \; of \; product, V = velocity \; of \; flow,$
$N = rotor \; speed \; and \; n = number \; of \; rotor \; pockets.$
The speed of rotation is controlled to regulate the material flow rate.

(c) **The Pipeline** is another extremely important part of the pneumatic system. The pipeline may be of seamless mild steel, stainless steel, aluminum, rubber and plastics which are wear resistant. Pipe bends should be avoided because of wear and corrosion and it is essential that flanges or couplings are used at bends to enhance maintainability and replacement. Pipeline diameter, wall thickness, and pipe materials are to be determined while selecting a pipeline. For $45°$ bends, the turning radius should be 6-8 times the pipeline diameter.

(d) **Disengaging and Collecting Devices** are usually in the form of a cyclone and fabric filter. In any pneumatic system, there is always trouble at the delivery end, in releasing the crop from the airstream and allowing the air to escape. The cyclones are widely used to assist separation of light materials by slowing down the material to minimize damage. Their choice depends on the amount of solid involved, the particle size range, the collecting efficiency required and the capital/running costs. In positive and negative pressure systems, an inlet filter or screen is used to remove dirt from the incoming air before it enters the blower.

10.3.4 Air-Activated or Air-Assisted or Fluidized Conveyor (Air slide)

The air slide conveyor system is an air-activated gravity type conveyor using low pressure air to aerate or fluidize the pulverized material to a degree which will permit it to flow on a slight incline by the force of gravity. The air velocity through the particles must be high enough to fluidize the particles for the solids to flow down the inclined channel. This system maintains an aerated state in the bulk solid once it is fed into the upper end of an inclined channel, by the continuous introduction of air (or any other gas) at a low rate through a false perforated bottom. Dry pulverized material e.g. flour, being fluidized by air, is being conveyed down slightly inclined $(4 - 10°)$ special trough by the gravitational force. Air pressure of 3 to 5×10^4 Pa is maintained in the air duct. Capacity of up to 1500 t/h (Butler, 1974) can be conveyed over a distance of 40m or more and with lesser (1/8 - 1/5) power requirements than mechanically-driven conveyors. The air is supplied by a fan or a rotary blower. The design fluidization velocity should be that at which further increase in velocity will not result in any significant increase in solid flow rate (Woodcock and Mason, 1987).

Two essential requirements are necessary for successful transportation of granular and powdery materials by the air slide. These are: -

(i)   That sufficient air flows into the material in the channel through the porous base to cause it to flow;

(ii)   That the downward slope is sufficient to permit a steady continuous flow of the fluidized powder (Fig. 10.5).

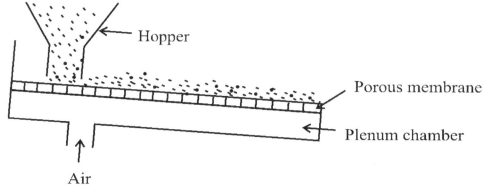

Fig. 10.5: Flow of aerated bulk solids in inclined channels

The conveying on the downward slope takes advantage of gravity to assist the flow. The main components of the air-gravity conveyor are shown in Fig. 10.6. The porous membrane is made of woven cotton, polyester, asbestos, plastic, ceramic tiles or laminated stainless steel mesh. One or two points supply air to the plenum chamber. Even though, air slides operated naturally as an open channel, covered ones ensure freedom from dust leakage but they are provided with vents through suitable filters. They are used widely for horizontal conveying and discharge at the bottom of silos, bunkers, storage bins, special railway wagons, truck-trailer transport and lorries (Hudson, 1954), enabling these containers have flat bases for greater capacity. They are used in stationary blow-tank-type conveying system. Another application of air slides is in the conveying of hot materials at temperatures of up to 135°C or even 530°C (Anon, 1954) when the porous membrane is made of ceramic tiles. A pneumatic tube package conveyor is used to convey unit loads at post offices, banks and departmental stores.

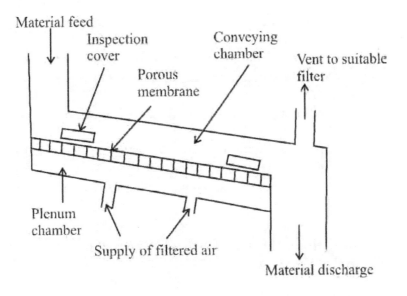

Fig. 10.6: Air - gravity conveyor

Advantages

The air-gravity conveyor has many advantages including that:

It is simple in construction and economical to use.

It has compactness, high capacity, and it is self-regulating.

It has low power consumption with trouble-free operation.

It can be made air-, gas-, or dust-tight.

It enjoys freedom from moving parts as well as those subject to rapid wear.

Disadvantages

The disadvantages of the air slide are few. They include:

Indispensable incline requiring certain floor area.

Not self-cleaning, but with blinding of pores in the top surface of the membrane.

10.3.5 Design Considerations for Pneumatic Conveyors

Design decisions, for pneumatic conveying, center upon the pipeline diameter and size and power of the air mover (Harris et al., 1965). Also, the designer thinks about methods of feeding the products into the pipeline to minimize pressure loss, and the type and sizes of the discharge units. However, for proper design of pneumatic conveyors, the following data are required:

–   Volume or mass throughput capacity i.e. mass flow rate.

–   Characteristics of the load (density, particle or lump size, etc.).

–   Rough layout of the conveying system, i.e. the length and configuration of the conveying path.

–   The type of system (i.e., positive or negative-pressure, dense or dilute-phase, etc.).

With these, the designer can now determine the following parameters of the conveyor: -

- The required conveying air velocity and volumetric air flow rate of the carrier air. (Srivastava et al., 1993; Marcus et al., 1990).
- The solids loading ratio or mass concentration (i.e. the ratio of the mass flow rate of conveyed material to the mass flow rate of the conveying air).
- The diameter of the pipeline
- The pressure-drop through the system arising from the resistance in the various sections of the pipeline.
- The specification of the air mover in terms of power, or flow rate and delivery pressure.

The steps for the determination of the above parameters of the conveyor are given below for the fan-driven system:

(i)   Required Conveying Velocity and Volumetric air flow rate

The minimum transport velocity (V_{cr}) in a pneumatic conveying system is influenced by many variables such as particle size, density and shape, cohesiveness and abrasiveness, and significantly by the mass concentration. For granular materials which cannot be conveyed in the dense-phase system but in a lean-phase system, the minimum transport velocity (V_{cr}) is around 16m/s. It could be higher for materials having large lumps with high density. Critical velocity (V_{cr}) is the lowest velocity at which the material does not accumulate and block the pipe and given by Eqn. 10.13.

$$V_{cr} = C_2\sqrt{(\mu agD)} \qquad\qquad \text{...10.13}$$

where

μ = mass concentration ratio
a = ratio of densities = $(\varrho_p - \varrho_f)/\varrho_f$ of load particles
D = pipe diameter
C_2 = empirical constant (Table 10.2)

Once V_{cr} has been determined, the required transport velocity (V) should be twenty percent higher for design purposes and to provide a margin of safety against pipeline blockage (Eqn. 10.14). Greater velocities will cause degradation of products, erosive wear on pipeline, increased power and filtration.

$$V = 1.20V_{cr} \qquad\qquad \text{...10.14}$$

The volumetric air flow rate is given by Eqn. 10.15.

$$V_o = (P_1T_o/P_oT_1)(\pi D^2V)/4 \qquad\qquad \text{...10.15}$$

where

P_1 and T_1 are pressure and absolute temperature at inlet, P_o and T_o are standard ambient pressure and absolute temperature. If P_o = 1 bar and T_o= 288 K, we get Eqn. 10.16.

Table 10.2: **Empirical Constant for Critical Velocity**

Materials	Constant C_2
Pulverized	0.1
Liable to slumping	0.25
Granular	0.3
Lumpy Materials (mc = 4%)	0.35

$$V_o = 226 P_1 D^2 V / (273 + t_1) \qquad m^3/s \qquad\qquad \dots 10.16$$

where

$P_1 = pressure\ (bar\ absolute);\ D = pipe\ diameter\ (m);$
$V = conveying\ air\ velocity\ (m/s);\ \ t_1 = temperature\ (^oC).$

For the air mover, the specified V_o will be larger by about 10-20% to take care of leakages. The volume of air required depends on the desired mass flow rate of the conveying air M_a.

(ii) **Mass Concentration or Solids Loading Ratio (Phase Density)**
The ratio of the mass throughput capacity (Q, t/h) of the system to the mass flow rate of the conveying air ($V_o \varrho_a$) called mass concentration or mass flow ratio (μ) is difficult to predict. However, it has the relationship shown in Eqn. 10.17.

$$\mu = Q/V_o \varrho_a = M_s/M_a \qquad\qquad \dots 10.17$$

where
M_s = mass of solids; M_a = mass of conveying air.

For dilute-phase system, μ ranges from 5 to 15. One can start with $\mu = 10$ and adjust this figure to match the predicted pressure-drop through the system.
Note that $\varrho_a = 1.2923\ kg/m^3 =$ density of air

In the negative pressure (fan-driven) system; $\mu = 1$; for high-vacuum, $\mu = 5$. In positive pressure systems, $\mu = 8 - 25$; for air-activated material flows, $\mu = 60 - 150$.

(iii) **Diameter of Pipeline**
Knowing the volumetric air flow rate (V_o) and the conveying air velocity (V), a suitable size of pipeline can be determined at the preliminary stage of the design using Eqn. 10.18.

$$D = (4 M_a / \pi \varrho_a V)^{\frac{1}{2}} \qquad\qquad \dots 10.18$$

where

M_a = mass flow rate of air; ϱ_a = density of air; V = conveying air velocity; D = diameter of pipeline. If $\mu = 10$; $M_s = 10M_a$, then Eqn. 10.18 becomes Eqn. 10.19.

$$D = (M_s/7.85\varrho_a V)^{\frac{1}{2}} \qquad\qquad …10.19$$

where

M_s = mass flow rate of solids. Note that ϱ_a and V should be estimated using anticipated delivery pressure (P) of the air mover as in Eqn. 10.20.

$$\therefore \varrho_a = P/RT = Px10^5/287(273 + t) \qquad kg/m^3 \qquad\qquad …10.20$$

where

$R = 287J/kg.K$ = characteristic gas constant.

The ratio of the material velocity (U) to conveying air velocity (V) also known as the index of relative velocity of solid particles (β) in the pneumatic system is given by Marcus et al. (1990) as:

$$\beta = \frac{U}{V} = 1 - 0.68d^{0.92}\varrho_p^{0.5}\varrho_a^{-0.2}D^{0.54}$$

$$\varrho_p = \mu\varrho_a V/U = \mu\varrho_a/\beta$$

where

d = particle mean diameter (m); D = pipe diameter (m); ϱ_p = particle density (kg/m³); ϱ_a = air density (kg/m³); μ = mass flow ratio. $\beta = 0.35 - 0.85$

(iv)   **Pressure-drop**
This is a difficult prediction to make for the pneumatic conveying system. However, computer-aided design techniques are now available (Busheall and Maskell, 1960; Leitzel and Morrisey, 1971). A review of some methods of predicting a pressure drop in a suspension flowing along a pipe is given by Woodcock and Mason (1987) as Eqn. 10.21.

$$P_s = P_g + P_p \qquad\qquad …10.21$$

where

P_s = total pressure drop in the suspension, P_g = pressure drop due to gas/air alone and P_p = additional pressure-drop due to solid particles.

For the dilute-phase conveying of agricultural products, it is acceptable to calculate the wall-friction pressure-drop for conveying velocities greater than 25m/sec. as Eqn. 10.22.

$$P_s = P_g(1 + \mu) \qquad \qquad \dots 10.22$$

where

μ = mass concentration or solid loading ratio. $\mu = 1$ for fan - driven system; $\mu = 5$ for high vacuum; $\mu = 8 - 25$ for air activated material flows.

According to Srivastava et al. (1996), the total pressure drop in a pneumatic conveying system is made up to six components given by Eqn. 10.23

$$P = P_1 + P_a + P_s + P_g + P_b + P_c \qquad \qquad \dots 10.23$$

where

P_1 = pressure drop due to air only = $\lambda_1(\varrho_a/2)V(L/D)$
P_a = pressure drop due to particle acceleration = $\mu V \varrho_a U$
P_s = pressure drop due to solids = $\mu \lambda_s(\varrho_a/2)V^2(L/D)$
P_g = pressure drop due to lift height = $\varrho_a g H$

P_b = pressure drop due to bends = $P_{ba} + P_{bs}$
P_{ba} = pressure drop at bend due to air = $L_{eq} = (K_c/\lambda_1)D$
P_{bs} = pressure drop at bend due to solid = $0.245\varrho_a V^2(M_s/\varrho_a V D^2)^{1.267}(R/D)^{-0.260}$

P_c = pressure drop in accessories = $0.7 - 1.4 kPa$ (Noyes and Pfieffer, 1985)
λ_l = air resistance factor = $0.0004 + 0.0313 R_e^{-0.32}$ (Marcus et al., 1990)
λ_s = solids friction factor = $0.0285\sqrt{(gD)}/U$ (Marcus et al., 1990)
R/D = ratio of bend radius and pipe diameter
K_c = fitting loss coefficient = 0.05 to 1.3 (ASHRAE, 1972)
R_e = Reynolds number = $\varrho_a V D/v$
v = viscosity of air (N.s/m^2)
H = lift height (m)
L_{eq} = equivalent length = length that produces the pressure drop in a straight pipe as that in the bend (m).
U = material or solid velocity (m/s).
V = conveying air velocity (m/s)
L = conveying length (m).

Having estimated the pressure-drop, the pipeline diameter must be selected. High pressure drops deserve larger diameters. When long pipelines require excessively high conveying velocity with the attendant product degradation or pipeline erosion, a stepped-pipeline is selected. This reduces both the conveying velocity and pressure drops along the line.

10.4 CALCULATION OF PNEUMATIC TRANSPORT SYSTEMS

In addition to the design considerations given above, pneumatic transport system parameters may also be calculated as follows:

(i)   The design of a pneumatic conveying system involves determining the conveying air velocity (dependent upon the size, shape and the density of the conveyed material), volume of conveying air (dependent on the desired mass flow rate), total pressure drop (sum of many pressure drops), and power requirements for the blower. Assuming that a = size of the largest lump, then the pipe diameter D > 3a, and the critical velocity as in Eqn. 10.13 is:-

$$V_{cr} = C_2\sqrt{(\mu a g D)}$$

where

C_2 = empirical constants (Table 10.2)

(ii)   The total required pressure (MPa) in the conveying system is given by Eqn. 10.24.

$$P = P_p + P_i + P_l + P_f \qquad\qquad ...10.24$$

where

P_p = pressure losses at the beginning of the pipeline
P_i = pressure loss due to inertia
P_l = pressure loss in short rising sections of the pipeline
P_f = pressure loss caused by the feeder(s)

(a)   The pressure losses at the beginning of the pipeline can be calculated using Eqn. 10.25

$$P_p = 0.1P_o(1 + C_3\mu a g \, D/V^2) \qquad\qquad ...10.25$$

where:

C_3 = 0.01 - 0.075 (lower values are taken for loads of higher density) and $V > V_{cr}$ by about 10 - 20%, is the actual conveying velocity. Also P_o = pressure loss of pure air determined by Eqn. 10.26 for isothermal flow.

$$P_o = 0.1\left[\sqrt{(M_a^2 RTfL_r/A^2 gDx10^8 + P_e^2/10^2)} - 1\right] \qquad\qquad ...10.26$$

where

M_a = mass flow rate of air (kg/s); R = 287J/kg.K. = universal gas constant; T = absolute temperature of surroundings; f = coefficient of pipe resistance (for 150 mm diameter pipe = 0.016 - 0.020, for 175 mm diameter = 0.015 - 0.018; for 200 mm diameter= 0.014 - 0.016); L_r = resolved pipe length (m); A = cross-sectional area (m^2); P_e = 0.015 MPa = pressure at end of pipeline.

Bends, valves and other fittings in the pipeline increase the frictional pressure loss and are considered as equivalent length of pipe, as in Section 10.3 above, of the same diameter as the main pipeline and added to the straight length of the pipeline to give the resolved pipe length.

For isochoric flow (i.e. constant volume and ΔP = constant), the pressure loss of pure air can also be calculated using Eqn. 10.27.

$$P_o = \xi V^2 / gD \qquad\qquad \ldots 10.27$$

where
ξ = coefficient of hydraulic resistance of pipe; P_o = standard ambient pressure.

(b)   The pressure losses due to inertia, P_{in} can be estimated from Eqn. 10.28.

$$P_{in} = (V^2 \varrho_g / 2)(1 + \beta\mu)10^{-5} \qquad MPa \qquad\qquad \ldots 10.28$$

where
ϱ_g = density of gas; β = index of relative velocity of solid particles (0.35 - 0.85) and for pulverized loads, β = 0.60 - 0.85.

(c)   The pressure losses (P_l) in short rising sections of the pipeline of height H_l (m) is got from Eqn. 10.29.

$$P_l = 10^{-5}(1 - \mu)\varrho_g g H_l \qquad\qquad \ldots 10.29$$

(d)   The pressure loss caused by the feeder (P_f) should be as small as possible so as not to interfere with the flow of product into the conveying line. Since the product feed is usually from atmospheric pressure to the conveying pipeline, air leakage must be minimized to prevent undue interference to the feeding process and avoid unnecessary reduction of the air available to transport the product. The common feeding devices have P_f = 0.01 - 0.8 bar. Other equations have been developed for calculating the various pressure drops to be accounted for in the design of pneumatic conveying systems. These are given in Marcus et al., (1990); Noyes and Pfieffer, (1985), and Srivastava et al., (1996), as shown above.

(iii)   The conveying air flow rate (m³/s) and the delivery pressure required at the blower or compressor discharge are two most important parameters for selecting the air mover. If V_o is the volumetric air flow rate required for conveying, the actual air flow rate (V_a) at the blower/compressor discharge is given by Eqn. 10.30.

$$V_a = kV_o \qquad\qquad \ldots 10.30$$

where

$k = 1.10 - 1.15$ is coefficient of air leakage in pipeline, feeder, etc.

(iv)   The required power of the blower/compressor motor is given by Eqn. 10.31.

$$HP_m = K_2 L_{bl} V_a / 1000\eta \quad kW \qquad \qquad ...10.31$$

where

$L_{bl}=$ theoretical work (Eqn. 10.32) of blower/compressor related to 1 m^3 of sucked-in air in isothermal compression (N-m/m^3), $K_2 = 1.1 - 1.2 =$ safety factor.

$$L_{bl} = 2303 P_e \, log(P_p/P_e) \qquad \qquad ...10.32$$

where

P_p and P_e are initial and final pressures respectively in compression and
$\eta = 0.65 - 0.85$ is blower/compressor efficiency.

However, the power requirement of the blower which depends on the conveying air volumetric flow rate (V_o) and the total system pressure drop (P) may also be calculated from the standard air equation (Eq. 10.32a) and corrected for temperature, humidity and altitude.

$$HP = PV_o/\eta_b \quad W \qquad \qquad ...10.32a$$

where

P in Pa; V_o in m^3/s and $\eta_b =$ blower efficiency (0.5 - 0.7).

10.5 HYDRAULIC/SLURRY CONVEYORS
10.5.1 Introduction

These conveyors, also known as slurry conveyors, convey products in suspension in a moving liquid. Slurry is a liquid, usually water, in which foreign material of varying quantities is suspended. Hydraulic transport as part of hydro-mechanization has been applied in the fields of mining, power plants, mineral processing, construction industry, water management, agriculture and chemical industry. In the past it has been used for high-tonnage, long distance transportation of solid materials (coal, copper, etc.) (Zandi, 1982; Duckworth et al., 1986). As shown in Fig. 10.7, the carrier liquid (water) enters the slurry tank into which the solids are poured to form the slurry. The main pump draws the slurry and delivers it over a pipe line to a destination where de-watering is done.

There exists considerable interest in the transportation, separation and handling of agricultural products by water either at harvest, packing points, etc. Also in agriculture, waste transport and disposal under safe conditions are other objectives of the system.

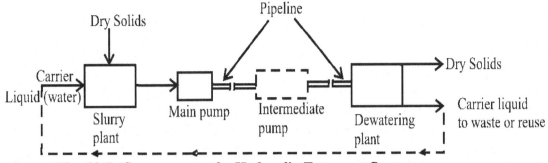

Fig. 10.7: Components of a Hydraulic Transport System

The water transportation can be achieved either by: -

i)   Fluming or sluicing the products gravitationally in open channels or sluice-ways at a slight downgrade (0.02 - 0.06).

ii)   Pumping the products and water through a pipe inclined up to 90° as well as on horizontal sections; or

iii)   Using both sluicing and pumping simultaneously.

Advantage of hydraulic transport

The advantages of hydraulic conveying are more apt for handling fruits and vegetables.

1.   There is the possibility of moving them over long distances; literally anywhere a pipeline can be placed. It moves at a high rate without any mechanical means along the path.

2.   For these agricultural products, mechanical damage is minimized; product cooling immediately after harvesting is achieved, thereby enhancing product shell life for good quality preservation.

3.   The system operates automatically (no problems in service) and enables conveyance to be combined with some processing operations such as cleaning or washing of easily perishable fruits and vegetables.

4.   The space requirement is low as no special equipment and no other infrastructure is needed.

5.   The system is environmentally friendly because it is a closed system with low labor requirement.

Disadvantages of hydraulic transport

The disadvantages of the hydraulic conveyor that limits its wide scale usage include:

1.   The nature of the load to be handled and the size of the particles which induce rapid wear of parts in contact with the slurry.

2.   There is comparatively high-power consumption at the pumps as well as high demand for transport water.

3.   De-watering of the mixture after transport may prove difficult and expensive.

10.5.2 Types of Mixtures

Hydraulic transport may deal with homogeneous mixtures (milk or salt) or with heterogeneous mixtures (tomatoes in water, or vegetables in saline). In the heterogeneous system, two or more phases are not interdependent and two or more substances are intermingled. The carrying liquid (water, oil, methanol, etc.) must not dissolve or affect adversely the product it is carrying. Each component retains its essential, original properties, with the product moving by means of either suspension or siltation.

In siltation the motion of particles is in a series of short intermittent jumps when the particles are large and the flow velocity is low. The rate will be proportional to the flow velocity and there is a deposit regime.

Suspension occurs when the particles are small (10^{-4} - 10^{-6} mm) and the flow velocity high. In this case the particles are carried by the fluid in some characteristic relative velocity - the particle speed is different from that of the liquid giving rise to a non-deposit regime. The transition from the deposit to the non-deposit regime is determined by the settling velocity which is an equivalent to terminal velocity in aerodynamic or pneumatic transport.

10.5.3 Settling Velocity (V_s) and Apparent Drag Coefficient ($C_D{}^1$)

As in pneumatic conveyors, these two parameters are very important in hydraulic conveyors. When a particle is dropped in a fluid, it travels downwards until a settling velocity is reached where the drag force and the buoyant force balance the weight of the particle. In lamina flow conditions and with particles of sphericity 0.85 to 1 in a viscous fluid, the drag coefficient is obtained with Eqn. 10.33.

$$C_D = 24/R_{ep} \qquad \qquad \dots 10.33$$

where

R_{ep}= particle Reynolds number; C_D = drag coefficient.

For $R_{ep}< 0.2$, we obtain V_s using Eqn. 10.34.

$$V_s = gd^2 \left(\varrho_p - \varrho_f\right)/18\varrho_f v \qquad \qquad \dots 10.34$$

where

V_s = settling velocity

For $0.2 < R_{ep} < 500$, the drag coefficient is given by Eqn. 10.35, while the settling velocity is given by Eqn. 10 36.

$$C_D = 18.5/R_{ep}^{0.6} \qquad \qquad \dots 10.35$$

$$V_s = 0.153(d^{1.14}/v^{0.43})\left[g\left(\varrho_p - \varrho_f\right)/\varrho_f\right]^{0.71} \qquad \qquad \dots 10.36$$

For $500 < R_{ep} < 2 \times 10^5$, the drag coefficient is as in Eqn. 10.37 and settling velocity in Eqn. 10.38.

$$C_D = 0.44 \qquad\qquad\qquad …10.37$$

$$V_s = 1.73\left[dg(\varrho_p - \varrho_f)/\varrho_f\right]^{0.5} \qquad\qquad …10.38$$

The development of a combination of dimensionless parameters as in pneumatic conveyors is also appropriate here to determine the settling velocity after estimating the particle's Reynolds number Re_p. For non-spherical irregularly shaped particles such as agricultural products, Hawksley's (1951) redefinition of drag coefficient Eqn. 10.10 and Reynolds number (Eqn.10.11) holds. However, use is made of equivalent or stokes diameter (d_e), thereby giving Eqn. 10.39 for settling velocity.

$$V_s = \left[4gd_e\phi_s(\varrho_p - \varrho_f)/3C_D\varrho_f\right]^{0.5} \qquad\qquad …10.39$$

Experience has shown that apparent drag coefficient (C_a) is one of the most important considerations in fluid flow for power calculations, and is given by Eqn. 10.40.

$$C_a^{0.5} = (C_D/\phi_s)^{0.5} \qquad\qquad …10.40$$

where

C_a = apparent drag coefficient and ϕ_s = sphericity or shape factor $1 \ge \phi_s \ge 0.4$

10.6   COMPONENTS OF A HYDRAULIC TRANSPORT SYSTEM
Four major components make up the hydraulic transport system for conveying slurries. These are the pump, the slurry preparation plant, the pipeline, and the de-watering equipment (Fig. 10.7).

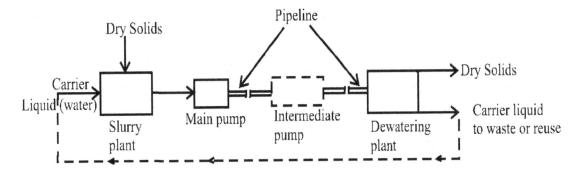

Fig. 10.7: Components of a Hydraulic Transport System

(a)   The Slurry Pump

This is the heart of the system. Many types are available for handling slurries (Thompson et al., 1972). The plunger and piston types or reciprocating pumps are used for finely dispersed and granular materials of size not greater than 2 mm. The roto-dynamic or centrifugal pumps are used for lumpy materials of size more than 100 mm. Table 10.3 shows the maximum values of the main factors that govern slurry pump selection: pressure, flow rate and nature of material. While plunger and piston pumps are similar in construction, the former is single-acting and the later could be either single or double-acting.

Table 10.3:   Performance of Slurry Pumps.

Type	Max pressure (bars)	Max flow (m³/h)	Max particle size (mm)	Material type	Mechanical efficiency (%)
Plunger	240-275	200	1.5	Abrasive	85-90
Piston	170-210	600	2	Non-abrasive	85-90
Centrifugal	40-50	11000	150	Coarse	40-75

(b)   Slurry Preparation Plant

The plant prepares the bulk solid by milling or grinding it to a size suitable for pumping, mixes it with the carrier liquid and introduces it into the conveying pipeline. Here, the slurry density and particle size both of which affect flow are adjusted in storage tanks - where screens are used to ensure that oversized particles do not enter the main pipeline. The size of the particles in the pipeline should not only give optimum flow characteristics but should be adequate for any subsequent processes.

i.   The Pipeline

The pipeline is the vein/artery of the pneumatic and hydraulic systems. It is made of mild steel, reinforced concrete, or high-density polyethylene (HDPE). The main problems with the pipeline are effects of corrosion and abrasive wear caused by cavitation and presence of suspended solids moving at high velocity. In HDPE pipeline, pipe bursts are possible due to high-pressures. Erosive or abrasive wear is usually combatted by lining the pipeline with rubber or plastics and using conveying velocities less than 3 m/s.

(d)   De-Watering Equipment

This helps to remove the carrier liquid. Three processes may be involved in de-watering: -

-     Sedimentation of the particle by gravity, assisted by centrifugal action;
-     Filtration of the carrier liquid which drains through a cake of the solid assisted by centrifugal action, pressure or vacuum.
-     Thermal drying at temperatures that will not affect adversely the product.

Using various forms of screens, rapping or vibration of the screens and washing the slurry over the screens with additional water and even holding the slurry in conical bottom settlement tanks, sedimentation processes take place. Separation of liquids/solids can also be done using the hydro-cyclone which is similar to the air cyclone. Centrifugal separation is done by using various forms of a centrifuge like the solid-bowl, screen-bowl or basket, or the rotating-bowl centrifuge. For details of how centrifugal separation functions check in Woodcock and Mason (1987). Recovery of fragile materials is better done by using vacuum and pressure filtration whose simplest form is the rotary drum filter.

10.7   CALCULATION OF HYDRAULIC TRANSPORT SYSTEMS.

The following data: the volume or mass throughput capacity, the characteristic of the load (density, particle or lump size, etc.), the length and configuration of the conveying path, help to determine the required speed of the carrier fluid, its flow rate, the diameter of pipeline, the resistance in various sections of the pipeline and the required head or pressure to overcome them and the power of the pumps or blowers. For heterogeneous mixtures, the maximum allowable speed must be such that mechanical injury to the product is avoided, while the minimum transport velocity should not allow settling of the products. The steps for calculating the required parameters of the hydraulic transport system are given below: -

10.7.1 Fine - dispersed Bulk Load

For fine-dispersed bulk loads (particle sizes < 0.2 mm) in turbulent flow:-

i.   The critical velocity is got using the following equation Eqn. 10.41.

$$V_{cr} = K\sqrt{(agD)} \qquad\qquad\qquad ...10.41$$

where

$K = 1.0 - 1.5$ = coefficient of mixture intermixing, D =   pipe diameter

$a = (\varrho_s - \varrho_w/\varrho_w)$= ratio of the densities of load and water

ii. The pipe diameter D should be checked to satisfy the condition given in Eqn. 10.42.

$$V = 4V_h/3600\pi D^2 \geq V_{cr} \qquad\qquad\qquad ...10.42$$

where

V_h = flow rate at hydraulic mixture (m³/h); V = speed or velocity of flow (m/s).

iii. The unit head loss (m/m) for the moving mixture is seen in Eqn. 10.43.

$$H_l = KH_o(1 + as) \qquad\qquad\qquad ...10.43$$

where

$H_o = K_a V^2/(gD)$ = unit head loss for a flow of pure water (m/m),

K_a = coefficient of hydraulic resistance of a smooth pipe,

s = volume concentration of hydraulic mixture (V/V_h),

a= ratio of densities of load and water.

10.7.2 For lumpy and granular materials (particle size > 2mm).

(i)   The critical velocity is got using the following Eqn. 10.44.

$$V_{cr} = C_1 \sqrt{(fagsD)} \qquad \qquad \dots 10.44$$

where

$C_1 = 8.5 - 9.5$ = experimental coefficient, f = generalized coefficient of material friction on pipe (0.10 - 0.75)

(ii)   The unit head loss for moving moisture has Eqn. 10.45.

$$H_l = H_o + fas \qquad \qquad \dots 10.45$$

where

s = volume concentration = 0.2 - 0.25

10.7.3 The Required Pressure Head

The required pressure head H_r (m) to overcome loss of head in both suction and delivery lines is given by Eqn. 10.46.

$$H_r = H_l + H_{lr} \qquad \qquad \dots 10.46$$

where

H_l= lifting head; H_{lr}= local head losses (5% of total losses in straight sections).

10.7.4. The Required Power

For a given throughput capacity V_o (m³/h), the required power of the pump motor (kW) is given by Eqn. 10.47.

$$HP = K_s H_r V_o \varrho_l / 367\eta \qquad \qquad \dots 10.47$$

where

$K_s = 1/\mu = 1.1 - 1.2$ is considered a safety factor

$\mu = 1/(1 + ß)$ and ß = ratio of water to solids by volume

$\beta = (\varrho_s - \varrho_w)/(\varrho - \varrho_w)$

$\eta = 0.7 - 0.9$ = efficiency of the pump

10.8   The Jet Pump (Fig.10.8)

The jet pump differs from all pumps by the absence of moving parts and its working member is liquid *per se* (Krivchenko, 1986). In Fig 10.8, the driving or head development liquid is supplied through the nozzle (2) while the solids-laden liquid to be transferred enters through pipe (1). They are forced at high velocity through the mixing tube or throat (3) into the diffuser (4). The driving liquid may be water while the transferred fluid may be solids-laden slurry. Due to friction and exchange of momentum at the surface of the driving fluid jet, the latter entrains the suction fluid from the slurry tank and moves it to the mixing chamber. The mixing chamber is where an exchange of momentum between the driving fluid and the suction fluid takes place. The diffuser converts the kinetic energy into potential energy and from there the mixed liquid flows into the discharge line.

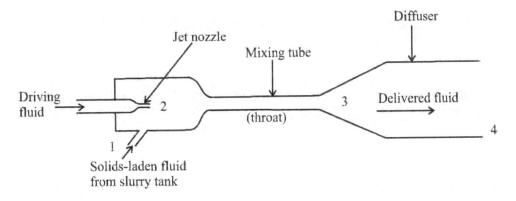

Fig. 10.8: Components of the Jet pump

The jet pump is a low-pressure high volume flow rate pump. Simplicity of design, absence of any moving parts, ability to handle muddy water, reliability, ruggedness, and low cost, more than compensate for the relatively poor efficiency of the pump (Saker and Hassan, 2013).

This pump moves the slurry on the principle of fluid velocity energy. The driving fluid moves through the jet nozzle at very high directional velocity onto the slurry from the tank. They mix at the tube or throat and the mixture is conveyed through the diffuser and into the delivery tube (pipeline). The flow rate of the driving fluid and the solids-laden fluid should be the same for reasonable delivery pressure.

Jet pumps may be used for pumping air by vacuum created in the chamber where (1) and (2) meet. However, the shape of the nozzle, the distance between the nozzle and the mixing chamber, the shape of the intake chamber and diffuser, must be strictly concentric. An error in selecting the distance between the nozzle exit section and the mixing chamber may result in impulse action of the jet which decreases its performance (Cherkassky, 1985).

The basic relationship is based on the conservation of energy. Consideration is given to the specific energy of flow in the different sections usually referred to the mass of the moving fluid, corresponding to the gravity of 1 N expressed in J/N or Nm/N. However, when the specific energy of liquid is referred to 1 kg of liquid flow, the specific energy is expressed in J/kg or m²/s² and called specific work or mass head. The specific power of the liquid flow can be expressed as N_s =

$\varrho g V H$ (J/s or W), where ϱ = flow density (kg/m³); g = acceleration due to gravity (m/s²); V = volumetric flow rate (m³/s) = Av; H = overall Bernoulli head = overall specific energy of flow (m); A = cross-section area of flow at section (m²) and v = fluid velocity at section (m/s).

Considering the different pressure points, the energy equation is given by Eqn. 10.48.

$$N_{s4} = N_{s1} + N_{s2} - F \qquad or$$
$$H_4 A_4 V_4 - H_1 A_1 V_1 + F = H_2 A_2 V_2 \qquad \qquad ...10.48$$

where

H = overall Bernoulli head; A = cross -sectional area
F = energy losses due to friction, turbulence and mixing at (3) in Fig. 10.8
H_2 = power or driving head;
V = Volumetric flow m³/s

The efficiency of the jet pump is determined by the ratio of the liquid useful energy to the energy input when losses are neglected (Eqn. 10.49).

$$\eta = (H_4 A_4 V_4 - H_1 A_1 V_1)/H_2 A_2 V_2 = (N_{s4} - N_{s1})/N_{s2} \qquad ...10.49$$

The values of $\eta = 0.27 - 0.35$ take into account only the losses involved in mixing of flows with different velocities (impact losses).

Advantages
The advantages of the jet pump include that:

- There is complete freedom of moving parts
- It is used to pump materials of sludge consistency e.g. fluid manure
- It has low initial cost
- It is simple in design
- It can move liquids and gases

Disadvantages
There are very few disadvantages. They include

- Very low pump efficiency
- Accurate dimensioning and careful manufacture are required

10.9   AIRLIFT PUMP (Fig. 10.9)
This operates with the same principles as the jet pump except that instead of having the driving liquid we have the driving air (1) pumped in at very high velocity through the jet (2). At the neck of the delivery tube (3) bubbles are generated which reduce the slurry density at the neck bringing about motion. With very high air pressure, wet materials are lifted into the delivery tube

(4) and transported.

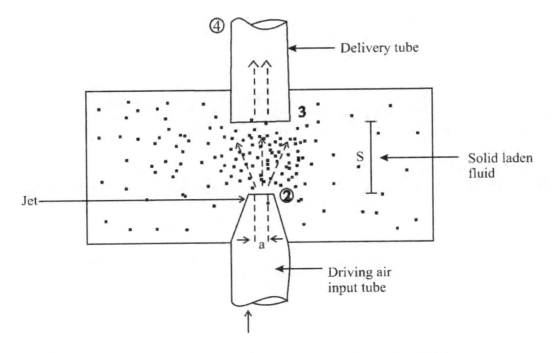

Fig. 10.9: *Airlift (Hydro-transport apparatus)*

An air lift pump is a device used to raise the liquid or slurries from a well or vessels through a vertical pipe, partially submerged in the liquid, by means of compressed air introduced into the pipe near the lower end (Clark and Dabolt, 1986; Abou Taleb and Abdallah, 2017). This type of lift is employed, for example, to deliver water and oil from drill holes. They are used to raise water from boreholes of any diameter and depth and are the most simple and reliable devices for this purpose. However, they are particularly important for raising water from boreholes of small diameters, where reciprocating and centrifugal pumps cannot be used. Also, they may be used in chemical and food industries where aggressive liquids have to be delivered to a small height. They may be used to raise liquids contaminated by sand, ash, or peat. In this case, airlift pump components contacted by the mixture should be of corrosion resistant materials or at least given a protective coat of paint or vanish.

Advantages
i.   The airlift is by construction simple.
ii.   It has no moving parts and is not affected by suspended particles.
iii. It is particularly advantageous for elevation of water from small diameter boreholes or wells.
iv.   It can be easily assembled at the construction site with air supplied it from a mobile compressor.

Disadvantages

The shortcomings of this airlift pump are:

    i.   There is the need for deep immersion of the pipe under the water level.
    ii.   It has comparatively low efficiency ($< 50\%$).
    iii.   It is not feasible for it to deliver liquid along horizontal or slightly inclined pipelines.
    iv.   There is the possibility of contamination of supplied liquid by compressor oil.
    v.   There is greatly increased oxygen content in the supplied liquid.

The Characteristics of the airlift pump

The characteristics of this apparatus are usually described by three ratios: Eqn. 10.50, Eqn. 10.51 and Eqn. 10.52.

$$R = Nozzle\ area/Throat\ area = A_2/A_3 \qquad\qquad ...10.50$$

For commercial production $0.25 < R < 0.625$.

$$M = Pumped\ capacity/Driving\ capacity = Q_3/Q_2 \qquad\qquad ...10.51$$

$$N = Net\ jet\ pump\ head/Net\ driving\ head = (H_4 - H_3)/(H_1 - H_4)...10.52$$

The efficiency of the airlift is given by Eqn. 10.53.

$$\eta = MxN = [Q_3(H_4 - H_3)]/[Q_2(H_1 - H_4)] \qquad\qquad ...10.53$$

This efficiency is normally about 35%. It is not above 40 - 50%, as regards energy of compressed air supplied, and amounts to only 15 - 25% as regards power consumption including compression losses (Krivchenko, 1986).

If an airlift pump delivers liquid of density ϱ (kg/m^3) at the rate of V (m^3/s) against a head of H (m), the pump useful output (J/s) is given by Eqn. 10.54.

$$N_{uf} = \varrho g V H \qquad\qquad ...10.54$$

Assuming that standard air under pressure P$_2$ (Pa) is fed to the airlift pump chamber at the rate of V$_1$ (m^3/s) to a pressure of P$_1$ (Pa). The power consumed by the compressor and expended in raising the liquid up the airlift pipe is got from Eqn. 10.55.

$$N_{com} = P_1 V_1 ln(P_2/P_1)(1/\eta_{ise}\eta_m) \qquad\qquad ...10.55$$

where

        $\eta_{ise} = isentropic\ efficiency;\ \eta_m = mechanical\ efficiency.$

The airlift pump efficiency in this case becomes (Eqn. 10.56).

$$\eta_{al} = \varrho gVH\eta_{ise}\eta_m/P_1V_1\ln(P_2/P_1) \qquad\qquad ...10.56$$

(This efficiency depends on the submergence and equals to about 0.5. It varies with V_1 and peaks before V reaches its maximum).

In designing the airlift, it is important to properly locate the throat (3) with respect to the nozzle (2). The distance separating the jet nozzle and the throat should be approximately equal to the nozzle diameter for an optimum configuration ($S \simeq d_2$). This limits us because of the size of material to be conveyed. In practice the separating distance is taken to be about 2 to 3 times the nozzle diameter [$S \simeq (2 - 3)\,d_2$].

10.10 SOLVED PROBLEMS

Q.10.1 Design a pneumatic conveyor to convey 25 t/h of wheat of density 0.95 t/m³ and sphericity 0.42 at a flow velocity of 13 m/s to a 4m high silo. Assume mass concentration ratio = 10 and initial pressure = 0.9 MPa.

Data: Given: $\varrho_p = 0.95\,t/m^3$; $Q = 25t/h$; $V = 13\,m/s$; *Product* = *wheat*
$\Phi s = 0.42$; $\varrho_a = 1.29\,kg/m^3$; $\mu = 10$; $P_p = 0.9MPa$; $H_1 = 4m$

Solution:
From Eqn. 10.14; $V = 1.2V_{cr}$; $V_{cr} = 13/1.2 = 10.83\,m/s$
From Table 10.1; $CD_1 = 0.10$.

The equivalent diameter of wheat is got using Eqn.10.10,

$$C_D^1 = \phi_s\, 4gd_e(\varrho_p - \varrho_a)/3\varrho_aV_{cr}^2$$

$$or\ d_e = 3C_D^1\varrho_aV_{cr}^2/4\phi_sg(\varrho_p - \varrho_a)$$

$$\therefore\ d_e = 3x0.1x1.29x10.83^2/4x0.42x9.81(950 - 1.29) = 45.39/16.48x948.71 = \mathbf{2.90mm}$$

The volumetric flow rate (Eqn. 10.17)

$$\mu = Q/V_o\varrho_a = 25/V_o x 1.29^{-03} = \mathbf{10}$$

$$\therefore V_o = \mathbf{1938\ m^3/h} \sim \mathbf{1940\ m^3/h\ or\ 32.3\ m^3/s}$$

The diameter of pipeline (Eqn.10.13)

$$V_{cr} = C_2\sqrt{(\mu a g D)} \quad (C_2 = 0.3\ is\ got\ from\ Table\ 10.2)$$

$$a = (\varrho_p - \varrho_a)/\varrho_a = (0.95 - 0.00129)/0.00129 = \mathbf{735.43}$$

$$\therefore D = V_{cr}^2/C_2^2 \mu a g = 10.83^2/0.3^2 x 10 x 735.43 x 9.81 = \mathbf{18.06\ mm}$$

But equivalent product diameter, d_e = 2.90mm
This shows that D = 6.228 d_e (which is adequate).

Total required pressure (Eqn.10.29) P = P_p + P_i + P_l + P_f with P_p = 0.9 MPa.

$$P_i = inertia\ pressure\ loss = (V^2/2)\varrho_a(1 + \beta\mu) x 1.0^{-5} MPa$$

Assuming ß = index of relative velocity of solid particles = 0.75

$$\therefore P_i = (13^2 x 1.29/2)(1 + 0.75 x 10) x 1.0^{-5} = \mathbf{0.0093\ MPa}$$

$$P_1 = pressure\ loss\ in\ pipe\ height = 1.0^{-5}(1 + \mu)\varrho_a g H$$

$$= 1.0^{-5}(1 + 10) x 1.29 x 9.81 x 4.0 = \mathbf{0.0056\ MPa}$$

Assume P_f = pressure loss in feeding device = **0.03 MPa.**

$$\therefore P = 0.9 + 0.0093 + 0.0056 + 0.03 = \mathbf{0.9449\ MPa}$$

The airflow rate at the blower (Eqn.10.30)

$$V_a = KV_o = 1.15 x 32.3 = \mathbf{35.145\ m^3/s}$$

The theoretical work of blower (Eqn.10.32)

$$L_{bl} = 2303 P_e log(P/P_e) = 2303 x 0.105 log(0.9449/0.105) = \mathbf{230.739\ Nm/m^3}$$

With a blower efficiency of 80%, the required power (kW) of the blower is given by Eqn.10.31 with K_2 = safety factor = 1.2

$$HP_{motor} = K_2 L_{bl} V_a/1000\eta = 1.2x230.739x37.145/1000x0.8 = \mathbf{12.86\ kW}$$

Q.10.2 The ratio of fresh pepper to water by volume in a hydraulic conveyor is 0.78. If the density of the pepper is 1.1 kg/1, how much pepper will be conveyed by a motor of 35 kW? Find the mixture density. If the density of the solids/water mixture is 1.2 kg/1, by how much will the throughput capacity increase for the same power and by how much will the power increase to maintain the new capacity at $\beta = 0.78$. Assume $\eta = 75\%$, and H = 10 m.

Data: *Given:* $\varrho_s = 1.1\ kg/cm^3$; $\eta = 75\%$; $HP = 35\ kW$;
$H = 10\ m$; $\beta = 0.78$; $\varrho = 1.2\ kg/cm^3$; $\varrho_w = 1.0\ kg/cm^3$

Solution:

Using Eqn. 10.47; $HP = \varrho_w V_o H/367\eta\mu$; $\mu = 1/1 + \beta = 1/1.78 = \mathbf{0.5618}$;
$But\ \beta = (\varrho_s - \varrho_w)/(\varrho - \varrho_w)$

Also; $35 = 1xV_o x10x1.78/367x0.75$ *and* $K_s = 1/\mu = \mathbf{1.78}$

V_o = *volumetric capacity of the pepper* = $\mathbf{541.22\ m^3/h}$
$\therefore \varrho = \mathbf{1.128\ kg/cm^3}$

If $\varrho = 1.2$, $\beta_1 = (1.1 - 1.0)/(1.2 - 1.0) = 0.5$ $\therefore \mu_1 = 1/1.5 = \mathbf{0.667}$

$HP = \varrho_w V_{o1} H/367\eta\mu_1$ $\therefore V_{o1} = HPx367x\eta x\mu_1/\varrho_w H = 35x367x0.75x0.667/10$
$= \mathbf{642.57\ m^3/h}$

$\Delta V = [(642.57 - 541.22)/541.22]x100\% = \mathbf{18.7\%}$

However, to maintain β=0.78 for the new V_{o1}, *the power required*:-

$$HP_{o1} = 1x642.57x1.78x10/367x0.75 = \mathbf{41.55kW}$$

$\therefore$ *power increase* $= \Delta HP = (41.55 - 35/35)x\ 100\% = \mathbf{18.7\%}$

The power and throughput capacity of the conveyor increased by the same percentage.

Q.10.3 Calculate the critical velocity and the power of the blower motor for a pneumatic conveyor. The initial data are: $\varrho_a = 1.29 \ kg/m^3$; $D = 0.8 \ m$; $a = 0.16$; $\mu = 90$; $P_p = 0.153 \ MPa$; $P_e = 0.105 \ MPa$; $V = 6.0 \ m/s$; $K = K_2 = 1.1 \ and \ C_2 = 0.3 \ and \ \eta = 80\%$; $\beta = 0.65$; $C_3 = 0.025$; $H = 15 \ m$; $P_f = 0.003 \ MPa$

Solution:

(**i**) $V_{cr} = C_2\sqrt{\mu agD} = 0.3\sqrt{90 x 0.16 x 9.81 x 0.8} = \mathbf{3.189 \ m/s}$

Since V > V$_{cr}$, the chosen diameter is adequate.

(ii) $V_o = air\ flowrate = AV = (\pi D^2/4)xV = (\pi x0.8^2/4)x6 =$ **3.016 m^3/s**

(iii) $V_a = KV_o = 1.1x3.016 =$ **3.318 m^3/s**

(iv) $L_{bl} = theoretical\ work\ of\ blower = 2303P_e log(P/P_e)$

But $P = P_p + P_i + P_l + P_f$

$P_i = (V^2/2)\varrho_a(1 + \beta\mu)x1.0^{-5} = (6^2/2)x1.29(1 + 0.65x90)x1.0^{-5} =$ **0.0138 MPa**

$P_l = 1.0^{-5}(1 + \mu)\varrho_a gH = 1.0^{-5}(1 + 90)x1.29x9.81x15 =$ **0.1727 MPa**

$\therefore P = 0.153 + 0.0138 + 0.1727 + 0.003 =$ **0.3425 MPa**

$L_{bl} = 2303x0.105 log(0.3425/0.105) =$ **124.17 Nm/m^3**

(v) $HP = K_2 L_{bl}V_a/1000\eta = 1.1x124.17x3.318/1000x0.8 =$ **0.5665 kW**

Q.10.4 Cassava flour of density 0.25 t/m³ is transported by 15.6 kg/s of air of density 1.29 x 10⁻³ t/m³ and discharged isothermally through a horizontal pipe of diameter 0.25 m at pressure 0.105 MPa and speed of 24 m/s. If the pressure loss in isothermal motion P_o = 0.106 MPa, calculate the critical velocity, total pressure and the required power of the motor of 85% efficiency. *Assume* P_{in} = 0.125 MPa, P_f = 0.012 MPa; C_2 = 0.25; C_3 = 0.075; $K_1 = K_2$ = 1.1; Q = 195 t/h. P_o = 0.106 MPa.

Data: ϱ = 0.25 t/m^3; ϱ_a = 1.29 x 10⁻³ t/m^3; Q_a = 15.6 kg/s; D = 0.25 m; V = 24 m/s; P_e = 0.105 MPa; P_{in} = 0.125 MPa; P_f = 0.012 MPa; Q = 195 t/h; C_2 = 0.25; C_3 = 0.075; $K_1 = K_2$ = 1.1; η = 0.85; P_o = 0.106 MPa;

Solution:
(i) *Mass concentration* = $\mu = Q/Q_a = 195x1.0^3/3600x15.6 =$ **3.47**
(ii) *Ratio of densities* = $a = (\varrho_s - \varrho_a)/\varrho_a =$
$(0.25 - 1.29x1.0^{-3})/1.29x1.0^{-3} =$ **192.8**

(iii) $V_{cr} = C_2\sqrt{\mu agD} = 0.25\sqrt{3.472x192.8x9.81x0.25} =$ **10.13 m/s**

(iv) $P_p = 0.1P_o(1 + C_3\mu ag\ D/V^2)$
(v)
$= 0.1x0.106(1 + [0.075x3.472x192.8x0.25x9.81/24^2]) =$ **0.013MPa**
For horizontal pipe, P$_l$ = 0

215

(v) $P = P_p + P_i + P_f = 0.013 + 0.125 + 0.012 = \mathbf{0.15\ MPa}$

(vi) $L_{bl} = 2303 P_e log(P/P_e) = 2303 x 0.105 log(0.15/0.105) = \mathbf{37.458\ Nm/m^3}$

(vii) $V_o = \pi D^2 V/4 = \pi x 0.25^2 x 24/4 = \mathbf{1.18\ m^3/s}$

$\therefore V_a = KV_o = 1.1 x 1.18 = \mathbf{1.30\ m^3/s}$

(viii) $HP_{bl} = K_2 L_{bl} V_a/1000\eta = \mathbf{1}.1 x 37.458 x 1.3/1000 x 0.85 = \mathbf{0.063\ kW}$

CHAPTER 10 Review Questions

10.1 Discuss briefly the main purpose of pneumatic and hydraulic conveyors. What agricultural products are frequently conveyed by pneumatic conveyors?

10.2 Define terminal velocity and settling velocity. Derive their expressions in terms of drag co-efficient. Define equivalent diameter, and state when it could be used.

10.3 What are the advantages and disadvantages of pneumatic and hydraulic conveyors? Describe the dilute phase in which agricultural products are moved.

10.4 How do pneumatic conveyors compare with other mechanical conveyors, e.g. augers, bucket conveyors, etc.? Why is the negative-pressure type not common in agricultural practice?

10.5 Name the three types of pneumatic conveying systems and describe any one of them. Which industries apply pneumatic conveyors?

10.6 What are the main components of pneumatic and hydraulic conveying system? What main factors govern slurry pump selection?

10.7 What are the advantages and disadvantages of air-gravity conveyors? What are the functions of the slurry preparation plant?

10.8 Enumerate the design considerations for pneumatic and hydraulic conveyors and what the designer is supposed to determine? Define mass flow ratio; equivalent length; critical velocity; settling velocity; apparent drag coefficient.

10.9 By what three ways can hydraulic conveying be achieved, and how is it used in agricultural engineering practice?

10.10 Describe how de-watering and the preparation processes are carried out in hydraulic conveyor system.

10.11 Describe a jet pump and an air-lift. What are their advantages and disadvantages? And where are they used in agricultural practice?

10.12 State the design conditions for optimizing efficiency in jet pumps.

10.13 Enumerate the six (6) pressure losses that make up the total system pressure drop in pneumatic conveyors.

10.14 Cassava flour of density 0.25 t/m^3 is transported pneumatically through a horizontal pipe of 0.65 m diameter at 44 m/s and pressure 0.105 MPa. Total system pressure is 0.25 MPa. What will be the blower power if $K_2 = 1.1$ and $\eta = 0.65$? *($V_o = 14.6$ m^3/s; $L_{bl} = 91.1$ Nm/m^3; $HP_b = 2.25$ kW)*

10.15 Grain of $\varrho_1 = 1.68$ t/m^3 is transported hydraulically with $\varrho_w = 1.05$ t/m^3 through $D = 500$ mm and $V = 300$ m^3/l; $V_h = 1200$ m^3/h. Calculate: (i) volume concentration of the hydraulic mixture; (ii) density of mixture; (iii) mass concentration of mixture; (iv) critical velocity of flow if $C_1 = 8.5$ and $f = 0.45$; (v) required HP for pump motor if $K_s = 1.2$; $H_r = 15$ m and $\eta = 0.65$. *($s = 0.25$; $\varrho = 3.57$ t/m^3; $\mu = 0.53$; $V_{cr} = 4.89$ m/s; $HP = 29.47$ kW).*

10.16 Tomato fruits of $\varrho = 2.4$ t/m^3 are transported by gravity using water of $\varrho_w = 1.0$ t/m^3 through a semi-circular chute of slope 21.8^0 and $L = 4000$ m and of hydraulic radius = 0.20. If Chezy coefficient = 50, find the stream velocity. If the open chute is to be rectangular with B/h = 3.5, find B and h for the same hydraulic radius. *($V = 14.14$ m/s; $B = 1.1$ m; $h = 0.314$ m)*

10.17 If the fruits in Q.15 are to be moved by 40 kg/s air of $\varrho_a = 1.2923 \times 10^{-3}$ t/m³ with $\mu = 96$; $C_2 = 0.3$, $C_3 = 0.095$ and discharged through a horizontal pipe of $D = 0.8$ m at $P_e = 0.105$ MPa in isothermal (27⁰ C) motion of V = 600 m/s, find the critical velocity, total pressure required and required motor power if $\eta = 0.80$ and $P_f = 0.012$ MPa. Take $\lambda = 0.015$, $R = 29.7$. *(V$_{cr}$ = 354.8 m/s; P = 0.148 MPa; L$_{bl}$ = 36.12 Nm/m³; V$_a$ = 331.78 m³/s; HP = 16.48 kW).*

10.18 In a pressure type hydraulic conveyor $a = 0.85$; $s = 0.55$; D = 0.20 m. If $C_1 = 9.0$, $f = 0.45$ and $V = 1.2V_{cr}$, determine the flow rate of the hydraulic mixture *(V$_{cr}$ = 5.78 m/s; V$_h$ = 784.73 m³/h).*

10.19 For a pressure type hydraulic conveyor, $\varrho_1 = 1.16$ t/m³; $V = 90$ m³/h; $f = 0.48$; $s = 0.21$, $H_r = 25$ m; $D = 0.9$ m and $\eta = 0.85$. Find the critical velocity of flow and the required power of the pump. Assume: $C_1 = 9.0$; $K_s = 1.15$. *(a = 0.16; V$_{cr}$ = 3.4 m/s; HP = 9.62 kW).*

11

GRAVITY ROLLER CONVEYOR

11.1 INTRODUCTION

This is one of the materials handling equipment used in conveying unit loads (boxes, bags, bails, packages etc.) and containerized high-volume materials with smooth surfaces which are sufficiently rigid to prevent sagging between rollers. Roller conveyors are a form of conveyor belt that utilizes rollers - evenly-spaced rotating cylinders - to allow objects to skate across its surface. They move material from one place to another destination, and often leverage gravity or implement small motors to do so. The transported material must have a rigid riding surface that is supported by a minimum of three of the rollers. Motion of the unit loads on spaced rollers is usually by gravity down an incline (Fig. 11.1). The principle involved is the control of motion due to gravity by interposing an antifriction trackage set at a definite grade or slope. For horizontal and upgrade conveyance the rollers are powered. The grade of fall or gradient or slope varies from 2 to 7 percent. The rolls fitted to a frame are so spaced that about three (3) rolls are always under each unit load. This conveyor is also used for conveying fresh fruits, e.g. tomatoes, apples etc. through a washing sprayer with the rollers closely spaced. The rolls are usually the same as idlers in belt conveyors.

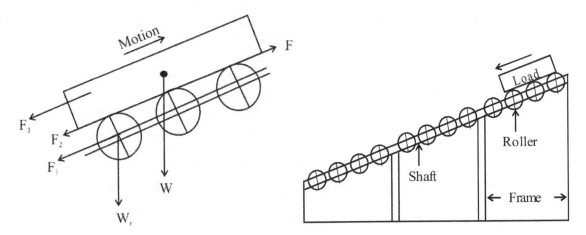

Fig. 11.1 Diagram of a gravity roller conveyor and its forces

Line-shaft roller conveyors are, as their name suggests, powered via a shaft beneath the rollers. These conveyors are suitable for light applications up to 20 kg such as cardboard boxes and tote boxes.

A single shaft runs below the rollers running the length of the conveyor. On the shaft there are a series of spools; one spool for each roller. A rubber o-ring runs from a spool on the powered shaft to each roller. When the shaft is powered the rubber o-ring acts as a chain between the spool and the roller making the roller rotate. The rotating of the rollers pushes the product along the conveyor. The shaft is usually driven by an electrical motor, which is generally controlled by an electronic PLC (Programmable Logic Controller). A PLC electronically controls how specific sections of the conveyor system interact with the products being conveyed.

Advantages

Advantages of this conveyor are

1. Quiet operation,
2. Easy of installation and maintenance,
3. Can be manufactured for almost all applications, at relatively low expense.

Disadvantage

A disadvantage of the roller line-shaft conveyor is that it can only be used to convey certain produce. Items conveyed have to be of adequate size and shape so that they cannot fall between the moving rollers.

There are other types of roller conveyors: gravity, belt-driven, chain-driven, and zero-pressure.

11.2   COMPONENTS OF THE GRAVITY ROLLER CONVEYOR

The main components of this conveyor include: -

(a)   **The rollers** which are of the same construction as those of the idlers in belt conveyors and are usually of steel or solid plastic or plastic-coated. Since the frictional resistance of the rollers influence the horsepower requirement, roller diameter, bearing design and roller spacing are important parameters to be considered in the design/selection of roller configuration. The diameter and capacity of rollers vary from 25.4 mm and 2.3 kg per roller to 108 mm and 816 kg per roller. They are fitted with antifriction bearings and seals to minimize frictional resistance that affect drive power and are mounted on pin journals, ranging in size from 15 mm to 35 mm. The spacing of the rollers varies with the size and weight of the materials or loads being conveyed. However, at least three rollers should be in contact with the load, at a time, to avoid hobbling.

(b)   **The drive system** for horizontal or up-grade conveyance includes the speed reduction mechanisms (reducers and gears), electric motors and controls, and safety devices. Motion is transmitted to the rollers from the drive system by a shaft carrying bevel gears (Fig. 11.2). This system powers the rollers and imparts motion to them which enable them to move the loads. Torque/speed control devices, such as fluid couplings, are widely used to allow some flexibility in the selection of electric motors.

(c)   **The mounting frame** carries the different sections of the conveyor. The sections measuring between 2 and 3 m long have a number of rollers fitted on them via the shaft. The frame is bolted down to avoid vibration and misalignment of the conveyor.

11.3   FORCES ACTING ON THE LOAD AND ROLLERS
The load being conveyed on rollers overcomes a resisting force which is a resultant of three horizontal force components (Fig. 11.1).

(a)   With the load resting on the rollers, there is friction of the roller bearings (F_1) referred to the outside roller diameter, given by Eqn. 11.1.

$$F_1 = (W + W_r i_1)f(d/D) \qquad \qquad \text{...11.1}$$

where
Wr = weight of rotating components of the roller (N);
i_1 = number of rollers under load;
f = coefficient of friction of roller bearing (Table 11.1);
d = diameter of roller pin journal (m);
D = outside roller diameter (m);
W = weight of load on the rollers (N).

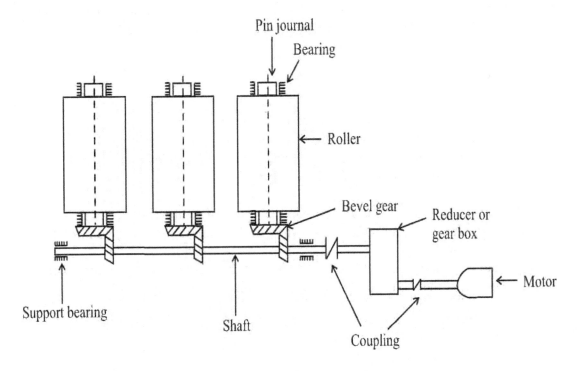

Fig. 11.2: Live-roll horizontal or upgrade conveyor

221

Table 11.1 Types of bearings and their coefficients of friction.

Bearing	Coefficient (f)
Ball/roller	0.015
Tapered	0.02
Open plain	0.10
Axle box	0.08

(b)   Before the load moves over the rollers, forcing the rollers to roll, there is a rolling resistance between the load and rollers to be overcome given by Eqn. 11.2.

$$F_2 = 2W\mu/D \qquad \qquad \text{...11.2}$$

where

$$\mu = coefficient\ of\ rolling\ friction = 0.03 - 0.12\ cm.$$

(c)   As the load slides over rollers, setting the rollers to acquire kinetic energy, there is sliding friction between the load and the kinetically driven rollers given by Eqn. 11.3.

$$F_3 = \psi W_r V_r^2 i/gL \qquad \qquad \text{...11.3}$$

where

$$\psi = roller\ mass\ concentration\ factor = 0.8 - 0.9;$$
$$Vr = rated\ circumferential\ speed\ of\ the\ roller\ (m/s);$$
$$i = number\ of\ rollers\ in\ the\ conveyor\ length, L;$$
$$g = acceleration\ due\ to\ gravity\ (m/s^2).$$

This resisting force (F_3) is an aggregate of those of all the rollers because by the time the load slides down the conveyor length, each of the rollers must have acquired kinetic energy.

The aggregate resistance force (F) to the motion of the load down a gravity roller conveyor is therefore the sum of the above forces as shown in Eqn. 11.4 and Eqn. 11.5.

$$F = F_1 + F_2 + F_3$$

$$= (W/D)(fd + 2\mu) + W_r(i_1 f\ d/D + \psi V_r^2 i/gL) \qquad \text{...11.4}$$

$$= WC_1 + W_r(i_1 C_2 + \psi V_r^2\ i/gL) \qquad \text{...11.5}$$

Where

$$C_1 = (fd + 2\mu)/D \qquad = \text{friction factor of load on conveyor}$$

$$C_2 = fd/D \qquad = \text{rolling friction factor of rollers}$$

It must be noted that before conveying, the rollers are at rest. As conveying starts, the load rolls and/or slides over the rollers imparting on them some circumferential speed due to a constant force of sliding friction. After the load had passed, the speed of the rollers gradually decreases to zero due to bearing friction unless another load comes on stream.

If the conveyor is inclined making an angle α with the horizontal (Fig. 11.3), the forces given in Eqn.11.1 through Eqn.11.5 assume the following forms shown in Eqn. 11.6 through Eqn. 11.10.

$$F_1 = (W Cos\,\alpha + W_r i_1)\,fd/D \qquad \qquad \ldots 11.6$$

$$F_2 = W(2\mu/D) Cos\,\alpha \qquad \qquad \ldots 11.7$$

$$F_3 = \psi(W_r V_r^2 i/gL) Cos\,\alpha \qquad \qquad \ldots 11.8$$

$$\therefore F = W(fd + 2\mu/D) Cos\,\alpha + \psi W_r V_r^2 i Cos\,\alpha/gL + W_r i_1 fd/D \ldots 11.9$$

$$= W C_1 Cos\,\alpha + \psi W_r V_r^2 i Cos\,\alpha/gL + W_r i_1 C_2 \qquad \ldots 11.10$$

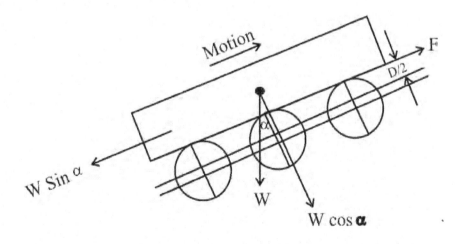

Fig. 11.3: Forces on an inclined conveyor

This aggregate force is for the load moving up or down the incline. It takes into consideration the frictional resisting and kinetic energy forces acting on both the load and the rollers.

When the load is being conveyed on an incline, a steady state motion is achieved when the moment of the propelling force equals the moment of the resisting force (Fig. 11.3) thus obeying the law of the conservation of momentum (Eqn. 11.11).

$$W Sin\ \alpha. D/2 = F. D/2 \qquad \qquad \dots 11.11$$

Dividing both sides of the equation by **Cos α** and transposing,

$$Tan\ \alpha = F/W Cos\ \alpha$$

And for small angle α, Cos α = 1, then Tan α = F/W making Eqn. 11.9 become Eqn. 11.12.

$$Tan\ \alpha = (1/D)(fd + 2\mu) + (W_r/W)(\psi V_r^2 i/gL + (fd/D)i_1) \qquad \dots 11.12$$

Eqn.11.12 gives the gradient of the incline or upgrade for steady state motion of a roller conveyor. It shows that the gradient depends on (a) type of load and (b) rate of conveying. The typical gradient ranges from 2° to 7° but can be increased to 12° to 14° for handling baled loads with soft uneven surfaces such as sacks of onions, mangoes, oranges, stockfish, etc.

11.4   POWER REQUIREMENTS OF ROLLER CONVEYORS

To transport unit loads on horizontal roller conveyors, power is absorbed in overcoming the frictional resistance encountered in moving the load. These include sliding and rolling resistance that occur between the load and the rollers along the length of the conveyor. If an upward movement is involved, power is absorbed in increasing potential energy of the load.

The power requirements for power-driven horizontal or inclined roller conveyors depend greatly on the weight of the load and the rollers, the capacity of the conveyor, the travelling speed of the load, and the length of the conveyor. This power requirement (kW) can be estimated using the formulae (Eqn. 11.13 and Eqn. 11.14).

For horizontal conveyor

$$N = ([C_1 QL/367] + [iW_r C_2 V/102])(1/\eta_d) \quad (kW) \qquad \dots 11.13$$

For inclined conveyor

$$N = ([QH/367] + [C_1 QL_h/367] + [iW_r C_2 V/102])(1/\eta_d) \qquad \dots 11.14$$

where
$\quad Q\ =\ throughput\ capacity\ of\ conveyor\ (t/h);$
$\quad L\ =\ conveying\ length\ (m);$
$\quad Lh\ =\ horizontal\ projection\ of\ conveyor\ length\ (m);$
$\quad H\ =\ lifting\ height\ (m);$
$\quad \eta_d\ =\ efficiency\ of\ driving\ mechanism\ (0.8-0.85).$

It is important to note that the power of the roller conveyor increases with an increase in those parameters upon which it depends as enumerated above.

11.5   MAINTENANCE OF ROLLER CONVEYORS

Maintenance is a prerequisite for safe operation and long life of a conveyor. In roller conveyors, wear takes place between the roller pins and the rollers, and between the loads and the rollers. In powered rollers, preventive maintenance should include checks on: -

(a)   The drive motor for unusual noise, alignment, lubrication, mounting bolts, bearings, motor fan and casing;

(b)   The reducer or gear box, also for unusual noise, alignment, vibration, oil level, oil change, leaking seals, mounting bolts;

(c)   The couplings and gears for tightness on shaft, keys and key-ways, alignment;

(d)   The rollers for free movement of rolls, even or excessive wear, mounting bolts, seals, bearings and lubrication;

(e)   The frame sections for proper alignment, the mounting bolts of the frames and the tightness of the frame connectors.

11.6 SOLVED PROBLEMS

Q.11.1   Eighty rollers of diameter ratio 0.1 and weight 20 kg rotating at 4.8 m/s are used to convey 3.4 t of tomato boxes over a length of 16 m at a load speed of 2.5 m/s. If a box rests on 3 rollers at a time, what will be the aggregate resistance to horizontal motion, if $\psi = 0.80, \mu = 0.10 \ cm., f = 0.015 \ and \ D = 150 \ mm$.

Data: $d/D = 0.1; \ i = 80, i_1 = 3; \ W_r = 20kg; \ V_r = 4.8 \ m/s; \ W = 3.4t;$
$L = 16 \ m; \ V = 2.5 \ m/s; \ \psi = 0.8; \ \mu = 0.1cm; \ f = 0.015;$

Solution: Using Eqn. 11.6; Eqn. 11.7; Eqn. 11.8

$$F_1 = (W + W_r i_1) \ (fd/D)g = (3400 + 20x3)x0.015x0.1x9.81 = \textbf{50.91 N}$$

$$F_2 = W.(2\mu/D)g = 3400x2x(0.1x10/150)x9.81 = \textbf{444.72 N}$$

$$F_3 = \psi W_r V_r ig/gL = 0.8x \ 20x4.8^2 x80/16 = \textbf{1843.2 N}$$

The total aggregate resistance force is given by:-

$$F = F_1 + F_2 + F_3 = 50.91 + 444.72 + 1843.2 = \textbf{2.34 kN}$$

Q.11.2.Calculate the power requirement for the conveyor in Q.11.1 above given $\eta_d = 0.80$. Assuming a steady-state motion, at what inclination will the conveyor above be installed to operate without power?

Data: as in Q.11.1; $\eta_d = 0.80$

Solution: Using Eqn. 11.13:-

$$N = ((C_1 QL/367) + (iW_r C_2 V/102))(1/\eta_d) \qquad (kW)$$

Where
$$C_1 = (fd + 2\mu)/D = 0.015x0.1 + 2x \ 0.1x10/150 = \textbf{0.0148}$$

$$C_2 = fd/D = 0.015x0.1 = 0.0015 = \textbf{1.5x10}^{-3}$$

$$Q = WV/L = 3400x2.5 \div 16 = \textbf{531.25kg/s}$$

$$N = (0.0148x531.25x16/367 + 80x20x0.0015x2.5/102)x(1/0.8) = \textbf{0.502 kW}$$

For steady-state motion and using Eqn. 11.12: -

$Tan\ \alpha = \frac{1}{D}(fd + 2\mu) + (W_r/W)([\psi V_r^2 i/gL] + [fdi_1/D])$

$= (1/150)(0.015x15 + 2x0.1x10) + (25/3400)([0.8x4.8^2x80/9.81x16] + 0.015x0.1x3)$

$= 0.0839$

$\therefore \alpha = 4.8^o$

Q.11.3 If the conveyor in Q.11.2 is inclined 7°, what will be the horse power of the motor if the throughput capacity and the load speed are to remain constant?

Data: $\alpha = 7^o$; $Q = 531.25\ kg/s, V = 2.5\ m/s$.

Solution: For inclined conveyor, using Eqn. 11.14: -

$N = ([QH/367] + [C_1QL_h/367] + [iW_rC_2V/102])(1/\eta_d)\ (kW)$

$H = 16\ Sin\ \alpha = $ **1.95 m**; $L_h = 16\ Cos\ \alpha = $ **15.88 m**

$\therefore N = ([531.25x1.95/367] + [0.0148x531.25x15.88/367] + [80x20x0.0015x2.5/102])x(1/0.8)$

$\qquad = \textbf{4.027}\ kW$

Q.11.4. A 120-roll roller conveyor takes pineapples up an incline through a water spray for washing. Find the μ and the HP required to convey 180 t/h of the pineapple at 2.5 m/s.

Data: $L = 25\ m$; $C_1 = 0.02$; $D = 50\ mm$; $W_r = 25\ kg$; $d/D = 0.15$; $\alpha = 5.6^o$; $V = 2.5\ m/s$; $Q = 180\ t/h$; $\eta = 0.85$; $f = 0.1$.

Solution: $Using\ C_2 = fd/D = 0.1x0.15 = o.015.\ But\ C_1 = C_2 + 2\mu/D$

$\therefore 0.02 = 0.015 + \mu/25$

$\therefore \mu = 0.125$

$Also, H = L\ Sin\ \alpha = 25Sin\ 5.6^o = \textbf{2.44m};$

$L_h = L\ Cos\ \alpha = 25\ Cos\ 5.6^o = \textbf{24.88m}$

Using Eqn. 11.14,

$$N = ([QH/367] + [C_1QL_h/367] + [iW_rC_2V/102])(1/\eta_d) \quad (kW)$$

$$\therefore N = ([180x2.44/367] + [0.02x180x24.88/367] + \\ [120x25x0.015x2.5/1000x102])(1/0.85) = \mathbf{1.7kW}$$

CHAPTER 11 Review Questions

11.1 What are the components of a roller gravity conveyor and what is this conveyor used for in agricultural engineering practice?

11.2 Mention the three horizontal force components this conveyor overcomes, and give their respective equations. What is the principle involved in gravity conveying?

11.3 Give the expressions for the load friction factor, C_1 and the rolling friction factor, C_2. For an inclined gravity conveyor, give the total force equation and the expression for the gradient for steady state motion.

11.4 Give the expressions for the power required for both a horizontal conveyor and an inclined roller conveyor.

11.5 What forces do the power of roller conveyors meant to absorb?

11.6 How would you maintain a roller conveyor? What factors influence the power requirements of gravity roller conveyors?

11.7 A roller conveyor 30 m long has 150 rollers each weighing 2 kg with a mass factor of 0.80. It transports 2.0 t of peeled cassava tubers up a 10^0 incline. *If $D = 0.2\,m, d = 0.02\,m$, $V_r = 2.5\,m/s$; $V_i = 0.8\,m/s$, $i_1 = 3, \mu = 0.80\,cm, f = 0.02, \eta_d = 0.85\ and\ \psi = 0.8$, what % increase in power will be attributed to 5% increase in incline if Q = constant? (Q = 192 t/h; C = 0.01; C_1 = 0.002; $HP_{10}{}^0$ = 3.4 kW; $HP_{10.5}{}^0$ = 3.6 kW; ΔHP = 4.7%)*

11.8 For a roller conveyor, $L = 12\,m, i = 60, i_1 = 3, \psi = 0.72, G_r = 2.4\,kg, d = 10\,mm, G = 6.5\,t, V_r = 2.0\,m/s, V_i = 0.45\,m/s, \alpha = 10^0, d/D = 0.1, \eta_d = 0.80, \mu = 0.06\,cm\ and\ f = 0.045$. What is the required motor power? *(Q = 877 t/h; C = 0.0165; G = 0.0045; HP = 6.8 kW)*

11.9 For a roller conveyor, $Q = 150\,t/h, H = 25\,m, i = 120, i_1 = 3, C = 0.07, G = 0.04, G_r/G = 1/1000, G_r = 2.5\,kg, V_r = 2.5\,m/s, V_i = 0.75\,m/s, \psi = 0.85$. *If $\eta_d = 0.90$,* tabulate the HP of the conveyor for $0 - 10^0$ incline in 2^0 increments. Assume a steady state motion and at what inclination will the conveyor operate without power? *(HP_o = 0.89 kW; HP_2 = 1.29 kW; HP_4 = 1.68 kW; HP_6 = 2.08 kW; HP_8 = 2.47 kW; HP_{10} = 2.85 kW; α = 4.16^0).*

11.10 For a roller conveyor, $G = 1.6\,t, \alpha = 10^0, L = 20\,m, i = 120, i_1 = 3, V_r = 2.8\,m/s, V_1 = 0.6\,m/s, G_r = 2\,kg, D = 0.16\,m, d = 0.016\,m, \mu = 0.08\,cm, f = 0.015, \eta_d = 0.85$. What resistance will be overcome by the roller conveyor system? For the conveyed material to move at doubled the speed, what % increase in power is required? *(F_1 = 23.3 N; F_2 = 154.6 N; F_3 = 78.8 N; F = 256.7 N; C = 0.0115; C_1 = 0.0015; Q = 172.8 t/h, Q_2 = 345.6 t/h; HP_1 = 2.05 kW; HP_2 = 4.10 kW; increase of 100%).*

11.11 For a powered roller conveyor, $L = 18\,m; i = 80; i_1 = 3; \psi = 0.75; G_r = 2.4\,kg; d = 10\,mm; G = 6.5\,t; V = 2.0\,m/s; V_t = 0.45\,m/s; \alpha = 12^0; d/D = 0.15; \eta = 0.80; \mu = 0.05\,cm; f = 0.045$. What is the total resistance and the required motor power of the conveyor. *(F_1 = 421.5 N; F_2 = 935.5 N; F_3 = 1.6 N; F = 1.4 kN; Q = 585 t/h; N = 8.08 kW)*

12

███████████████████████████████████████

FANS AND BLOWERS

12.1 INTRODUCTION

Fans and blowers are the major components of an airflow system to supply or exhaust air or flue gas to meet the needs of a variety of systems. Fans are used for building heating, ventilation and cooling for a wide variety of equipment. The fan/blower increases the pressure of a flow stream to offset the pressure losses that result from system resistance (Gamble, 1996). They are used in agricultural processing in connection with drying, ventilating, heating, cooling, refrigeration, aspirating, cleaning, elevating and conveying. While these are machines that move air/gases, pure or mixed with small solid particles, they differ from each other by their pressure ratios, ε, i.e. the ratio of the outlet pressure (P_2) to the inlet pressure (P_1).

Fans have a pressure ratio of up to $\varepsilon < 1.15$ at 1.2 kg/m^3 flow density.
Blowers operate at $\varepsilon > 1.15$ without artificial cooling
Compressors operate at $\varepsilon > 1.15$ and artificially cooled.

With $\varepsilon > 1$, it is evident that $P_2 > P_1$ and thus these machines impart energy to the air or gas thereby increasing their output pressures. Usually, fans, blowers and compressors are used interchangeably. However, compressors operate at pressures $\geq for\ 6879\ N/m^2$ with an intake pressure below atmospheric and a discharge pressure which is atmospheric or slightly higher. Table 12.1 shows the different parameters of various air moving machines.

Table 12.1 Main Parameters of Different Air-Moving Machines.

Type	Function	Capacity Q m³/min.	Compression ratio, ε	Rotating speed, rpm
Reciprocating	Compressors	0 – 500	2.5 - 1,000	100 - 3,000
Rotatory	Blowers	0 - 500	1.1 - 3	300 - 15,000
	Compressors	0 – 500	3 – 12	300 - 15,000
Centrifugal	Fans	0 - 6,000	1 - 1.15	300 - 3000
	Blowers	0 - 5,000	1.1 - 4	300 - 3,000
	Compressors	100 - 4,000	3 – 20	1,5000 - 45,000
Axial	Fans	50 - 10,000	1 - 1.04	750 - 10,000
	Compressors	100 - 15,000	2 – 20	500 - 10,000

Blowers operate at either low pressure ($< 1000 \text{ N/m}^2$) or medium pressure ($1000 - 3000$ N/m^2). They are commonly used to convey by imparting enough kinetic energy to the material to carry it through the conveying pipe to the desired height (Srivastava *et al.*, 1996). For the transport of hay, grain and straw products, a high pressure$\geq for \; 3000 \; N/m^2$) is used. Fans operate at pressures $< 6879 \text{ N/m}^2$ and may include centrifugal fans, axial fans, blowers or exhausters.

In dealing with **compressors**, consideration is given to the heat of compression and variation in the specific weight of the air or gas. For fans and fan blowers, these are of minor importance and are usually neglected.

Since these machines play an important role in agricultural engineering, not only in elevating and conveying, consideration is hereby given to their selection, adaptation and design. However, compressors are employed in agricultural processing as packaged units (as in refrigeration) and would not be given detailed analysis in this book. As for fans and blowers which are installation specific, the agricultural engineer or processor must select, adapt or design them for each individual situation. The selection of these machines is one of the most important decisions to be made during the design of a pneumatic conveyor or other systems in which they are used. Because of increasing costs, this largest single item of capital expenditure, in the system in which it is used, must be carefully designed and selected since the potential conveying capacity of the conveyor is dependent upon the correct choice being made. The air movers available for pneumatic conveying applications range from fans and blowers with high volumetric flow rates at relatively low-pressure ratios and low rotating speeds (Table 12.1) to positive displacement compressors (reciprocating or rotary screw machines) with higher pressures, higher volumetric capacities and higher rotating speeds for long distance conveying and vacuum pumps.

In general, both the supply pressure required and the volumetric flow rate to be delivered, give the rating of the air-mover. The supplied pressure must be sufficient to overcome the operating pressure-drop over the conveying line, and additional losses in connectors, feeding devices and filter units. The volumetric capacity depends upon the conveying velocity and size of the pipeline.

12.2 FANS

Fans provide high volumetric flow rates at low pressures and are used for dilute-phase conveying as in agricultural processing systems. They may be provided to control the flow of air through the equipment or through a series of components interconnected within a system as in pneumatic conveyors. In these systems use is made of the following types of fans: -

(a)   Axial-flow or propeller fans in which air flow is parallel to the shaft or axis. They have two or more blades of sheet or air-foil shape, narrow or wide, warped or smooth and may have uniform or varied pitch. They have some driving mechanism supports for either belt or direct connection. Other models include tube-axial, and vane-axial fans.

(b)   Centrifugal or radial-flow fans which have impellers which turn the parallel incoming air through 90^0 and to be discharged radially. They have only one moving part, the fan rotor weldment. There are single inlet and double inlet types. There is a pressure increase due to the work of the centrifugal force moving the air-gas from the shaft center towards the

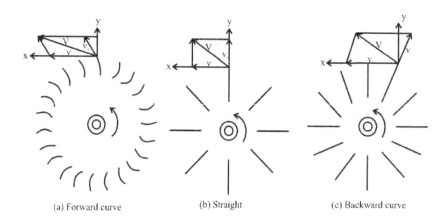

Fig. 12.1: Three types of centrifugal fan rotors, with velocity diagrams

impeller blade tip or periphery. The difference in types of centrifugal fans arises from the difference in blade curvature and number (Fig.12.1). Their performance characteristics are highly dependent on the type of blade used: radial tip (forward curved) (Fig. 12.1a), straight (Fig. 12.1b), and backward tip (Fig.12.1c).

Advantages of different types of centrifugal fan blades
The advantages of different types of centrifugal fan blades are: -

(i)   **Radial Tip (Forward Curved) Blades** have the advantages of none overloading horsepower, high capacity of size, excellent abrasion resistance, very stable operation, and essentially self-cleaning capabilities. But they are not as efficient as backward curved blades. They are used for erosive gas applications.

(ii)   **Straight Blades (or Radial Blades)** have the advantages of good abrasion resistance, simplified maintenance (particularly with blade replacement), and wide range of capacities. They have low operating efficiency and an overloading horsepower characteristic. They have limited applications in agricultural practice.

(iii)   **Backward Curved Blades** (air-foil type) have the advantages of higher efficiency (> 90%), very stable operation, low noise level, ideal capabilities for high-speed service, and none overloading horsepower characteristics. They should not be used where high levels of large or abrasive particles are present. The air-foil blades are used for almost all force draft and induced draft centrifugal fan applications.

12.2.1 Axial Fans
These fans consist of impellers having cantilever vanes fixed to the hub (Fig.12.2) at an angle to the plane of rotation making the flow somewhat twisted in the process of turning the rotor. Because the rotating impeller does not move in an axial direction, it moves the air/gas along the axis. The flow and pressure outputs are controlled by adjusting the pitch of the fan blades using either a mechanical or a hydraulic mechanism. The fan blades are supported on the fan bub with anti-friction thrust bearings. Using lightweight fan blade materials (aluminium or magnesium)

protected with corrosion resistant coatings reduces fan hub strength and bearing thrust load requirements thereby reducing axial fan cost significantly.

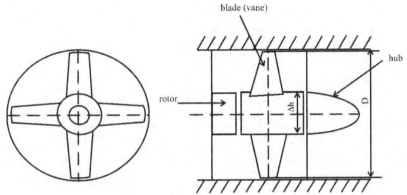

Fig. 12.2: Four - vane axial flow fan

The following basic theoretical considerations are given in relation to axial-flow fans.

(a)   The continuity equation (Eqn. 12.1) gives the inlet (1) and the discharge (2) throughput capacities as equal.

$$\varrho_1 A_1 C_1 = \varrho_2 A_2 C_2 \qquad (kg/s) \qquad\qquad\qquad …12.1$$

where

$\varrho = air/gas\ density\ kg/m^3$; A = duct cross-sectional area (m²);
C = axial component of absolute velocity (m/s).

(b)   The energy equation (Eqn. 12.2) gives the conversion of energy from kinetic to potential that takes place during the operation of axial fans. No energy is added to the flow through the impeller of an axial flow fan. There is dissipation of energy because of friction and vortex formation in the flow passage, leakage through clearances, mechanical friction in bearings and seals.

$$(W_1^2/2) - (W_2^2/2) = \int_1^2 (dP/\varrho) + \Delta L \qquad\qquad …12.2$$

where

W = relative velocity m/s; P = pressure N/m²; ϱ = density kg/m³; L = specific energy transforming into heat (Nm/kg).

(c)   The moment equation is used to calculate the forces of interaction between the flow through the vanes of an axial-flow fan.

233

12.2.2 Design Considerations for Axial-Flow Fans

The following parameters are considered when designing, selecting or adapting axial-flow fans: -

- – Main dimensions of the fan
- – The pressure head developed
- – The shaft power
- – The air horsepower
- – The efficiency

(a) **The main dimensions** are determined from both Euler's and continuity equations. However, H = head (m) of air to be moved by the fan; V_s = capacity (m³/s); and the physical constants of the air/gas e.g. density must be specified. When the fan is coupled directly to an electric motor it assumes the rotational speed of the motor. However, the fan's peripheral velocity must not exceed 100 m/s to avoid loud noise.

The hub diameter ratio (Eqn. 12.3) is taken as:

$$v = D_h/D_{out} = 0.4 - 0.8 \qquad\qquad\qquad …12.3$$

and the impeller diameter determined from Eqn. 12.4.

$$D_{out} = 2.9(V_s/K_o nv(1 - v^2))^{⅓} \qquad\qquad …12.4$$

where

D_{out} = outer diameter of fan cross-sectional area (m); v = hub diameter ratio; V_s = volumetric capacity (m³/s); n = rotating speed (rpm); K_o = velocity ratio

$$K_o = C_a/U_h = \quad axial\ absolute\ velocity\ (m/s)/tangential\ (peripheral)velocity\ (m/s)$$

Once v and K_o are given, Eqn. 12.4 determines the impeller diameter which is seen to decrease with increasing n making it advisable to use high rotational speeds for fans. Also, greater v is selected for high - pressure fans. From Eqn. 12.3, the hub diameter (Eqn. 12.5) is given as:

$$D_h = vD_{out} \qquad\qquad\qquad\qquad …12.5$$

and vane length (Eqn. 12.6) $\quad l_v = (D_{out} - D_h)/2 \qquad …12.6$

(b)**The head or pressure** against which the fan operates must be known for proper design or selection of the fan. The aim of the design is to obtain a uniform stream from the fan. When air is forced back through the fan near the hub (Fig. 12.3), re-circulation or turbulence occurs with lowered efficiency. This is eliminated or reduced by: -

(i) Warping of blades,
(ii) Large streamlined hub

(iii) Cylindrical housing
(iv) Adding straightening vanes behind the rotor

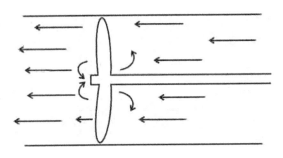

Fig. 12.3: Re-circulation

Reduction of turbulence reduces noise and improves efficiency. Forcing air through a duct system causes loss in pressure (called system resistance). The loss is both frictional (ΔP_f, at the walls of the duct) and dynamic (ΔP_d, at sudden enlargements and contractions and at changes in direction). The dynamic losses are also called velocity pressure losses. These losses are given in Eqns. 12.7a and 12.7b.

$$\Delta P_f = f \varrho V^2 L / gD \quad (kg/m^2) \qquad \qquad ...12.7a$$
$$\Delta P_d = \varrho k V^2 / 2g \quad (kg/m^2) \qquad \qquad ...12.7b$$

  where
  $\Delta P = pressure\ loss\ (kg/m^2)$, $\varrho = gas\ density\ (kg/m^3)$, V = gas/air velocity (m/s), L = duct length (m), D = duct diameter (m), g = acceleration due to gravity (m/s^2), f = dimensionless friction factor based on relative roughness and Reynold's number, k = friction constant based on duct system geometry.

By adding the friction losses to the dynamic losses, the total system resistance is determined. From the friction and dynamic losses equations, the system resistance changes directly as the gas density and is directly proportional to the square of the gas velocity (or volumetric flow rate or average duct velocity) for a set of flow conditions. With unchanged system geometry ($f, L, D, g\ and\ k$) only $\varrho\ and\ V$ influence system resistance and can be predicted for any flow condition once the pressure is known.

Using Euler's equation and assuming that the drift velocity is the same at inlet and discharge, the discharge coefficient

$$\psi = C_a / U$$

  where
  C_a = axial absolute velocity and U = drift velocity.

This coefficient determines the volumetric capacity per unit area of the vane cascade, thus the theoretical head given by Eqn. 12.8.

$$H_{th} = (U/g)C_a(ctg\ \beta_1 - ctg\ \beta_2)$$

$$= (U^2/g)\psi(ctg\ \beta_1 - ctg\ \beta_2) \qquad\qquad ...12.8$$

and the theoretical pressure as Eqn. 12.9.

$$P_{th} = \varrho g H_{th} = \varrho U^2 \psi(ctg\ \beta_1 - ctg\ \beta_2) \qquad\qquad ...12.9$$

> where
> $\beta_1\ and\ \beta_2 = entry\ and\ discharge\ angles;\ \varrho = density\ of\ air\ or\ gas$

The actual pressure (Eqn. 12.10) takes into consideration the cascade efficiency, η_c.

$$\therefore P = \eta_c P_{th} = \varrho U^2 \psi(ctg\ \beta_1 - ctg\ \beta_2) \qquad\qquad ...12.10$$

> where
> η_c ranges 0.90 - 0.94

(c)   **The power** input on the impeller via the electric motor is the shaft power on which the capacity of the fan depends. Also, the rotational speed of the fan is assumed to be equal to that of the motor. However, the air horsepower (Eqn. 12.11) is the output power of the fan.

$$Air\ horsepower = V_s \varrho H g/1000 \qquad (kW) \qquad\qquad ...12.11$$

> where
> V_s = volumetric capacity (m³/s); $\varrho = density\ (kg/m^3);\ H$ = total head (m)

The shaft horsepower as emphasized above is given by Eqn. 12.12.

$$Shaft\ power = QgH/1000\eta \qquad (kW) \qquad\qquad ...12.12$$

> where
> Q = mass flow rate (kg/s); $\eta = total\ efficiency\ (0.7 - 0.9)$

(d)   **Efficiency** is the ratio of air horsepower to shaft horsepower. There are energy losses in axial flow fans because of friction and vortex formation in the flow passage, leakage through clearances, mechanical friction in bearings and seals. These have given rise to hydraulic efficiency (η_h) which depends on the design of the suction approach and discharge portion as well as the mechanical efficiency (η_m) due to friction in seals, bearings and disc. The leakage loss is governed by the volumetric efficiency (η_{vol}). And the gross or total efficiency (Eqn. 12.13) is the product of the three efficiencies.

$$\eta = \eta_h \; x \; \eta_m \; x \; \eta_{vol} \qquad \qquad \ldots 12.13$$

The hydraulic efficiency is given by Eqn. 12.14

$$\eta_h = P_{st}/P_{th} = (H_{th} - h)/H_{th} = h/H + h \qquad \qquad \ldots 12.14$$

where
$$P_{st} = \text{static pressure (N/m}^2\text{)}, \; h = \text{head loss (m)}.$$

This efficiency ranges from 0.75 to 0.96 and is affected by (a) the passage configuration (b) finish of inner surfaces and (c) viscosity of the liquid.

The mechanical efficiency is given by Eqn. 12.15.

$$\eta_m = HP_{out}/HP_{in} \qquad \qquad \ldots 12.15$$

It ranges from 0.92 to 0.98 and is affected by (a) mechanical characteristics of the fan, (b) fan design and (c) in-service conditions of machine-bearing.

The volumetric efficiency is given by Eqn. 12.16.

$$\eta_{vol} = V_s/(V_s + \Delta V_s) \qquad \qquad \ldots 12.16$$

where
$$\Delta V_s = leakage \; losses \; (m^3/s)$$

It ranges from 0.96 to 0.98 and it is affected by the amount of radial clearance. The leakage loss is usually insignificant and this efficiency can be neglected. However, the range of values for η_m and η_{vol} show that axial fans maintain high efficiencies.

12.2.3 Axial fan characteristics

These characteristics relate head (the total and static pressures), shaft power, and gross efficiency to the volumetric capacity of the fan at a specific constant rotational speed. For different rotating speeds, these characteristics are recalculated according to similarity relations.

In Fig. 12.4, the power-curve approximates a horizontal line but actually decreases with increasing V_s. This makes it imperative to start the fan under a load, i.e. slide valves on the discharge pipe should be open. When the fan is throttled, the power requirement increases, owing to greater resistance head. The power unit must provide this extra power otherwise it may be overloaded and result in difficulties.

The pressure variation is saddle-shaped with a steady portion of this characteristic curve on the right of peak B. The efficiency curve has a pronounced peak. Variation in the operating conditions can cause abrupt changes (for rigidly fixed vanes) in efficiency leading to lower efficiency; and for adjustable vanes that can turn during operation, there is no appreciable loss of efficiency.

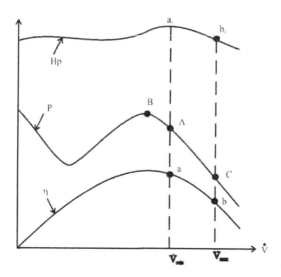

Fig. 12.4: Axial - flow fan characteristics (HP=horsepower; P=static pressure; η=efficiency)

The line a - a_1 (Fig. 12.4) gives the maximum permissible pressure at A to be 0.9 that at B. This allows the efficiency to be reduced to about 0.9 of maximum efficiency by line b - b_1.

12.2.4 Capacity Regulation

In applications that require fans to maintain variable and controlled pressure or flow conditions, the fans are equipped with pressure or flow regulating capabilities. This ensures the fan delivers a specific discharge V_s (m^3 /s) into the associated piping system according to a prearranged schedule. This causes the fan parameters (P, HP, η) vary as shown in Fig. 12.4.

Fan capacity can be adjusted by the following methods: -

(i)   **Speed Regulation** in which a variable-speed motor or constant-speed motor with a speed variant (hydraulic coupling) is inserted between the motor shaft and the fan. Though an expensive method with its complicated fan units, there is no power loss due to hydraulic resistance and therefore it is adjudged the most efficient method.

(ii)   **Throttling** is accomplished by adding extra-hydraulic resistance to the fan piping system i.e. by closing the throttle control. This method is particularly unfavorable economically because the power either remains constant or grows as the capacity is reduced by this method. The greater the amount of regulation, the higher the inefficient power consumption per unit volume of the fluid.

(iii) **Adjustable impeller vanes** or inlet guide vanes or adjustable pitch blade control, when used to regulate capacity reduces power intake and is therefore considered very economical.

For any type of fan, the worst method of capacity regulation is by throttling because of its greatest power consumption. The best method is by speed regulation either using hydraulic coupling or thyristor frequency converters.

12.2.5 Centrifugal or Radial - Flow Fans

These are the fans normally used in pneumatic conveying applications. Centrifugal fans are also widely used in industry and municipal engineering to ventilate buildings, aerate work stations, draw off dangerous gases generated in factories. They utilize the radial-flow impellers which are the principal component of the centrifugal machine. The impeller directs the incoming air through 90^0 due to centrifugal force which discharges the air towards the periphery of the impeller. The energy imparted through the profiled vanes of the impeller increases the pressure in the flow. The pressure-producing capability of radial- flow fans vary with blade depth, tip speed and blade angles. Centrifugal fan blades are mounted on an impeller (or rotor) that rotates within a spiral housing (shroud).

Fig. 12.5 shows a sectional view of the main part of the impeller. The main shroud is carried on the shaft via a hub. The suction approach to the impeller has diameter D_o. At the point where the blade becomes vertical is the inlet diameter D_1 . At the blade tip is diameter D_2 which is surrounded by the outlet passage.

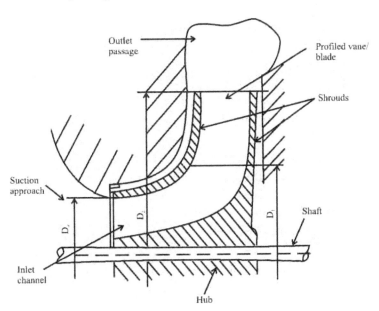

Fig. 12.5: Impeller of centrifugal machine

Cherkassky (1985b) stated that a characteristic design parameter of the centrifugal fans is the ratio D_2/D_1 between the outlet and inlet diameters of the impeller blade channels. In standard design this ratio is within 1.20 and 2.55 with the best fans having the ratio of 1.42 to 1.67; and the blade radial length is (0.084 - 0.16) D_2. Blade widths (b_1 and b_2) also relate to D_2/D_1 in the following manner given by Eqn. 12.17.

$$b_2/b_1 = D_2/D_1 \qquad \qquad \dots 12.17$$

So, properly selected diameter ratios are important when adjusting fan capacity by means of guide vanes.

(a)   For centrifugal fans, the energy equation (Eqn. 12.18) relates the relative motion of an incompressible, inviscid liquid in rotating channels.

$$\therefore (P_1/\varrho) + (W_1^2/2) - R_1^2(\omega^2/2) = (P_2/\varrho) + (W_2^2/2) - R_2^2(\omega^2/2) \qquad \dots 12.18$$

The difference in pressure is given by Eqn. 12.19

$$P_2 - P_1 = (\varrho/2)(W_1^2 - W_2^2) + (\varrho/2)(R_2^2\omega^2 - R_1^2\omega^2) \qquad \dots 12.19$$

Eqn.12.19 shows that the pressure built up is made up by the kinetic energy of relative motion and the work of centrifugal forces. Using the energy conservation law, the energy balance of flow through the impeller vanes can be written as Eqn. 12.20.

$$L_{th} = ((P_2 - P_1)/\varrho) + ((C_2^2 - C_1^2)/2) + gh \qquad \dots 12.20$$
     where

> L_{th} = specified energy imparted to the gas in the impeller channel and h = head losses in impeller channels.

Thus, the energy added to flow do three things:
    i.   Builds up the stream pressure
    ii   Increases the flow kinetic energy, and
    iii   Is expended in overcoming resistance

The L_{th} can be written in terms of the theoretical head, H_{th} as in Eqn. 12.21

$$L_{th} = gH_{th} \qquad \qquad \dots 12.21$$

(c)   The difference between the moments of per second momentum of the discharged gas V_s of density ϱ at the inlet and outlet sections of the impeller is used to determine the torque applied to the gas. This is given as Eqn. 12.22.

$$M_{th}\Delta t = (\varrho V_s C_2 r_2 - \varrho V_s C_1 r_1)\Delta t \qquad\qquad \dots 12.22$$

where

C = absolute velocity, M_{th} = theoretical torque.

But $r_1 = R_1 \cos\alpha$ and $r_2 = R_2 \cos\alpha$; where α = angle between peripheral and absolute velocities.

(d) The power imparted (Eqn. 12.23) to the flow in the impeller vanes is: -

$$HP_{th} = \varrho V_s L_{th}/1000 \qquad (kW) \qquad\qquad \dots 12.23$$

It has been said that most of the energy imparted by the fan is kinetic energy so the static efficiency is introduced to measure the static head developed by fans. The total (gross) and static (regain) efficiencies are given by Eqn. 12.24 and Eqn. 12.25, respectively.

$$Gross\ (total)\ efficiency\ \eta = \varrho V_s g H/1000 HP \qquad \dots 12.24$$
$$Static\ (regain)\ efficiency\ \eta_{st} = \varrho V_s g H_{th}/1000 HP \qquad \dots 12.25$$

The ratio η_{st}/η varies for different fan types according to the blade exit *angle β.* η_{st} *is about $20 - 30\%$ less than η*

However, the power of the fan drive motor (Eqn. 12.26) takes into account the following:
-

  –   Possible deviation from design conditions of operation
  –   Reduction in fan efficiency
  –   Deterioration of motor insulation during prolonged service.

$$HP_{motor} = k\varrho V_s H g/1000\eta\eta_{tr} \qquad\qquad \dots 12.26$$

where

k = power reserve factor = 1.05 to 1.2; η_{tr} = transmission efficiency: (for direct motor connection η_{tr} =1; for V-belt connection η_{tr} = 0.92)

(d) The shaft power/energy is greater than the useful energy given to the gas because some are wasted as hydraulic, mechanical and leakage losses. As in axial fans (Section 12.2.2), the hydraulic losses emanate from friction and vortex formation. As in Eqn.12.14: -

$$Hydraulic\ efficiency\ \eta_h = H/(H + h) = 1 - (h/H_{th})$$

Since there is impeller-to-casing clearance, some gases are leaked into the suction side of the fan causing **leakage losses,** ΔV_s (see Fig. 12.6). This makes the discharge through the impeller to be $(V_s + \Delta V_s)$. As in Eqn.12.16, the volumetric efficiency is given by: -

$$\eta_{vol} = V_s/(V_s + \Delta V_s)$$

The power generated inside the fan by the impeller vanes acting on the flow is the indicated power given by Eqn. 12.27.

$$HP_i = \varrho(V_s + \Delta V_s)g(H + h) \qquad \qquad ...12.27$$

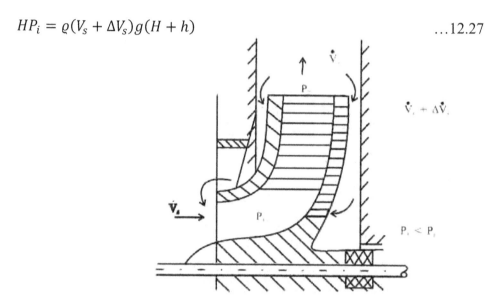

Fig. 12.6: Leakage losses in centrifugal fan

Thus, the vane efficiency, which takes into account only the leakage and hydraulic losses and not friction losses, is taken as in Eqn. 12.28.

$$\eta_i = \varrho V_s gH/[\varrho g(V_s + \Delta V_s)(H + h)] \qquad \qquad ...12.28$$

The above Eqn. 12.28 can be written as Eqn. 12.29.

$$\eta_i = \eta_{vol} \times \eta_h \qquad \qquad ...12.29$$

This gives the indicated power shown in Eqn. 12.30.

$$HP_i = QgH/\eta_{vol}\eta_h = HP_{usf}/\eta_i \qquad \qquad ...12.30$$

where
HP_{usf} = useful power.

Because of mechanical friction in bearings, stuffing boxes (shaft seals) and between the outer surfaces of the impeller and gas, gross mechanical efficiency is given in Eqn. 12.31.

$$\eta_m = HP_i/HP_{motor} \qquad \qquad ...12.31$$

This efficiency is affected by the design and in-service condition of the fan bearings and shaft seals and the finishing of the impeller non-active surfaces. This makes the motor horsepower to be as given in Eqn. 12.32.

$$HP_{motor} = HP_i/\eta_m = HP_{usf}/\eta_i\eta_m = QgH/\eta_m\eta_{vol}\eta_h = QgH/\eta ...12.32$$

where
$$\eta = \eta_m\eta_{vol}\eta_h$$

12.2.6 Characteristics of Centrifugal Fans

As in axial fan, these characteristics reflect plots of the head, (H and H_{st}), the shaft power (HP_{motor}) and the efficiencies (η and η_{st}) against the volumetric capacity (V_s) at a constant rotating speed and air density of 1.2 kg/m^3 derived from standard conditions. When actual conditions exist only the pressure and the shaft power vary in proportion to the conveyed gas density as given in the expression below.

$$i.e. \quad P = P_o \varrho/1.2 \ and \ HP_m = HP_o \varrho/1.2$$

Some characteristics are given in Fig. 12.7 for gas of density 1.2 kg/m^3, shaft speed 800 rpm and discharge outlet diameter 500 mm. It must be emphasized that fan characteristic curves depend for their shape on the diameter ratio D_2/D_1, blade exit angle ß, and the blade profile.

For variable-speed characteristic curves, the plots are constructed in compliance with similarity relations derived from geometric (relating to equal angles and permanent relations between sizes etc.), kinetic (relating to permanent velocity ratios at corresponding points) and dynamics (relating to permanent force ratios at corresponding points) similitudes. These similarity relations help the designer to design a series of geometrically similar types of fans in which the model fan is geometrically similar to the full-size fan. However, fan performance characteristics are developed from test data illustrated on flow vs. static pressure curves (Fig. 12.7).

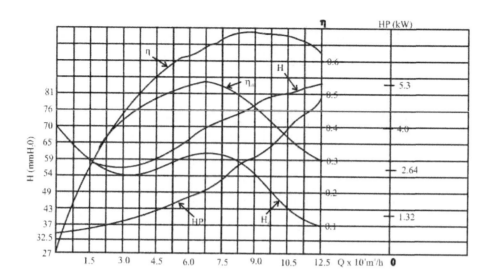

Fig. 12.7: Characteristics of centrifugal fan at n = 800 rpm

12.2.7 Fan Laws

In addition to static pressure capability, fan power requirements (brake horsepower) or fan efficiency (ratio of air horsepower to the actual shaft brake horsepower) are shown on fan performance curves. Fan performance curves are developed by testing model fans and the results used as a basis for determining the performance capabilities of full-size geometrically similar fans on the basic relationships between fan size, fan speed and air density. These relationships are generally called **Fan Laws**. With the following fan parameters: diameter size (D, m), velocity or rotating speed (n, rps), volumetric flow (V_s, m^3/s), static pressure (P_s, m), air density (ϱ, kg/m^3), and horsepower (HP, kW), head (H, m), t_{ip} speed (T_{ps}, m/s) and efficiency (η), three main laws are established.

Fan Law A

This law states that as the fan size (D) increases, the flow (V) increases as the cube of the size ratio and increases directly as the velocity (n) ratio. Model and full-size fan efficiencies must be the same for this Fan Law A.

$$Fan\ Flow \quad V_{s1} = V_{s2}(D_1/D_2)^3(n_1/n_2)$$

Fan Law B

This law states that as the gas density (ϱ) increases, the pressure (P) increases linearly, and as the gas velocity (n) increases, the pressure (P) increases with the square of n ratio and also the square of the size (D) ratio. This is dependent on the gas density which affects velocity and fan pressures.

$$Fan\ Pressure \quad P_{s1} = P_{s2}(D_1/D_2)^2(n_1/n_2)^2(\varrho_1/\varrho_2)$$

244

Fan Law C

Fan brake horsepower is a product of Fan Law A (Fan Flow) and Fan Law B (Fan Pressure). It is also affected by density change which makes the power ratio equal to the product of the diameter ratio to the 5th power, the speed ratio to the 3rd power and the density ratio to the 1st power.

$$Fan\ Power\ HP_1 = HP_2(D_1/D_2)^5(n_1/n_2)^3(\varrho_1/\varrho_2)\ or\ HP_1 = V_1P_1$$

The above relationships are true for fans 1 and 2 being geometrically similar and their efficiencies η_1 and η_2 are equal. When fan laws are being used to predict fan performance for various operating conditions and fan sizes, changes in fan size must be geometrical while their efficiencies must be the same. And in fan design these factors are used to derive the affinity (fan) laws.

12.2.8 Fan (Affinity) Laws

These laws relate the performance variables for any homologous series of fans and apply them to both axial- and radial-flow types based on incompressible flow. They help manufacturers and designers to build a series of geometrically similar types rather than arbitrary types of different sizes and geometry. In determining the relationships between the various parameters, the following variables are considered: fan size (discharge diameter), rotating speed, gas/air density, volumetric capacity, static/total head (pressure), fan power input, total efficiency, and sound-power level (for noise control). An important quantity that determines the similitude of flows is the specific speed. The specific speed (n_s) of a given fan is defined as the number of revolutions required to deliver 1 m^3/s of gas at a conventional pressure of 30 kgf/m^2 or 294 N/m^2 under best efficiency conditions and given in Eqn. 12.33.

$$n_s = 3.65nV_s^{\frac{1}{2}}H^{\frac{3}{4}} \qquad\qquad ...12.33$$

> where
> n = fan speed (rpm); V_s = specific flow rate (m^3/s); H = head (N/m^2)

If the fan designer considers fan capacity and fan pressure as the independent variables, the relationships for specific diameter, specific speed and specific sound power level are given in Eqn. 12.34 as:-

$$D_s = DH^{\frac{1}{4}}/V_s^{\frac{1}{2}}; \quad n_s = nV_s^{\frac{1}{2}}/H^{\frac{3}{4}};$$
$$L_{ws} = L_w - 10log\ V_s - 20log\ P_t \qquad\qquad ...12.34$$

> where
> $H = P_t/\varrho$

Proper use of the following fan law relationships will result in equal efficiency between operating points P_{t1} and P_{t2}.

Fan Law 1: This law (Eqn. 12.35) applies to change in speed ($n_1 < n_2$) for a given fan handling air at a given density. Efficiency, size and density do not change $\eta_{t1} = \eta_{t2}$, $D_1 = D_2$, and $\varrho_1 = \varrho_2$ (Fig. 12.8).

$$V_{s1}/V_{s2} = n_1/n_2 \; ; \; P_{s1}/P_{s2} = (n_1/n_2)^2; HP_1/HP_2 = (n_1/n_2)^3 \quad \ldots 12.35$$

Since the fan diameter and the air density do not change, the law shows the ratio of volumetric flow rate equal to the speed ratio, while the static pressure ratio and the horsepower ratio are directly proportional to the square and cube of the speed ratios respectively. The Eqn. 12.35 defines the constant efficiency operating line for alternative operating speeds.

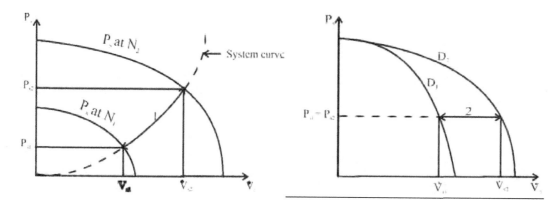

Fig. 12.8: Fan Laws 1 and 2

Fan Laws 2 and 3: These account for changes in fan performance due to proportional changes in fan size (D). They are based on either constant tip speed or constant n but without change in air density.

Fan Law 2: This law (Eqn. 12.36) deals with change in D ($D_1 < D_2$) with constant tip speed ($T_{ps1} = T_{ps2}$) and density ($\varrho_1 = \varrho_2$) (Fig. 12.8).

$$n_1/n_2 = D_2/D_1 \; ; \; V_{s1}/V_{s2} = HP_1/HP_2 = (D_1/D_2)^2 \qquad \ldots 12.36$$

Here, speed ratio equals diameter ratio. However, the volume ratio and the power ratio are equal and are each equal to the square of the diameter ratio for the same tip speed and same air density.

Fan Law 3: This law (Eqn. 12.37 deals with change in D ($D_1 < D_2$) with constant n ($n_1 = n_2$) and same density ϱ ($\varrho_1 = \varrho_2$). This law is used to generate performance data for geometrically proportioned families of fans (Fig.12.9).

$$T_{ps1}/T_{ps2} = D_1/D_2 \; ; \; P_{s1}/P_{s2} = (D_1/D_2)^2;$$

$$V_{s1}/V_{s2} = (D_1/D_2)^3 \; ; \; HP_1/HP_2 = (D_1/D_2)^5 \qquad \ldots 12.37$$

246

In this Law 3, the ratio of the tip speeds equals the diameter ratio. However, the pressure ratio, the volume ratio and the power ratio are respectively equal to the square, cube and fifth power of the diameter ratio. Eqn. 12.36 and Eqn. 12.37 define the constant efficiency operating line for alternative fan sizes.

Fan Laws 4, 5 and 6 apply to situations when the density ϱ changes ($\varrho_1 > \varrho_2$)

Fan Law 4: This law (Eqn. 12.38) deals with constant volume system ($V_{s1} = V_{s2}$), fan size ($D_1 = D_2$) and speed ($n_1 = n_2$) (Fig. 12.10).

$$P_{s1}/P_{s2} = \varrho_1/\varrho_2 = HP_1/HP_2 \qquad \qquad \ldots 12.38$$

The ratios of pressure, density and horsepower are all equal to one another

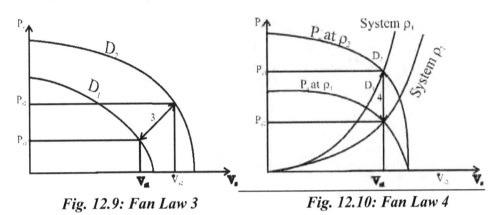

| *Fig. 12.9: Fan Law 3* | *Fig. 12.10: Fan Law 4* |

Fan Law 5: This law (Eqn. 12.39) deals with constant static pressure system ($P_{s1} = P_{s2}$) and fan size ($D_1 = D_2$) (Fig. 12.11).

$$V_{s1}/V_{s2} = n_1/n_2 = HP_1/HP_2 = (\varrho_2/\varrho_1)^{\frac{1}{2}} \qquad \qquad \ldots 12.39$$

The volume ratio, speed ratio and power ratio are equal and are each equal to the square root of the inverse of the density ratio.

Fan Law 6: This law (Eqn. 12.40) deals with constant mass flow system ($Q_1 = Q_2$) and fixed fan size ($D_1 = D_2$) (Fig. 12.11).

$$V_{s1}/V_{s2} = n_1/n_2 = P_{s1}/P_{s2} = \varrho_2/\varrho_1 \; ; and$$

$$HP_1/HP_2 = (\varrho_2/\varrho_1)^2 \qquad \qquad \ldots 12.40$$

Here, the ratios of volume, speed, static pressure are all equal to the inverse of the density ratio, while the horsepower ratio equals the square of the inverse of the density ratio.

Fan laws 4 and 6 are the basis for selecting fans for other than standard air density using catalogue fan tables based in standard air. Combination of the parameters related in the fan laws will help the designer to study experimental data and use them to study similar fans to be designed. Some of these parameters are given below.

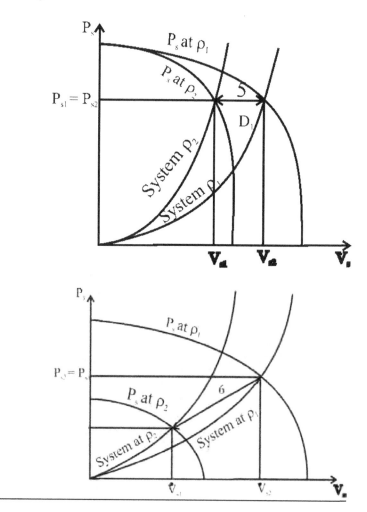

Fig. 12.11: Fan Laws 5 and 6

FAN CALCULATIONS

(1)   **Volumetric Capacity**
The volumetric capacity of the fan is calculated using Eqn. 12.41.

$$V_{s1}/V_{s2} = (D_1/D_2)^3(n_1/n_2)(\eta_{vol1}/\eta_{vol2}) \qquad ...12.41$$

If the fan sizes are the same ($D_1 = D_2$), the volumetric capacity ratio is given in Eqn. 12.42.

$$V_{s1}/V_{s2} = (n_1/n_2)(\eta_{vol1}/\eta_{vol2}) \qquad \text{...12.42}$$

(2) Total Head

The ratio of the total head of the fan is calculated with Eqn. 12.43.

$$H_1/H_2 = (D_1/D_2)^2(n_1/n_2)^2(\eta_{h1}/\eta_{h2}) \qquad \text{...12.43}$$

With $D_1 = D_2$, the total head ratio is given by Eqn. 12.44.

$$H_1/H_2 = (n_1/n_2)^2(\eta_{h1}/\eta_{h2}) \qquad \text{...12.44}$$

(3) Pressure Developed

The ratio of the pressure developed by the fan is given by Eqn. 12.45.

$$P_1/P_2 = (\varrho_1/\varrho_2)(H_1/H_2) = (D_1/D_2)^2(n_1/n_2)^2(\varrho_1\eta_{h1}/\varrho_2\eta_{h2}) \quad \text{...12.45}$$

For constant fan size ($D_1 = D_2$), the pressure ratio is shown in Eqn. 12.46.

$$P_1/P_2 = (n_1/n_2)^2(\varrho_1\eta_{h1}/\varrho_2\eta_{h2}) \qquad \text{...12.46}$$

(4) Shaft Power

The shaft power ratio is calculated from Eqn. 12.47.

$$HP_1/HP_2 = (\varrho_1 V_{s1}\, gH_1\eta_2/\varrho_2 V_{s2}gH_2\eta_1) = (D_1/D_2)^5(n_1/n_2)^3(\varrho_1\eta_2/\varrho_2\eta_1) \qquad \text{...12.47}$$

If the fans differ slightly in size and speed of rotation, $\eta_1 = \eta_2$. However, for accurate calculations, it should be noted that efficiency sometimes increases with an increase in fan size, and the performance of the fans is determined not only by its characteristics but also by those of the associated piping system. The similarity relations of fans when the parameters of V_s, $H, P, and\ HP\ vary\ with\ D, n\ and\ \varrho$ are summarized in Table 12.2.

12.2.9 Fan Selection

The factors affecting fan selection are as follows: -

1. The quantity of air to be moved per unit time (V_s) is determined by the size and type of installation;

2. The estimated system resistance and variations involving the total resistance head or static pressure (P_{st} and H);

3. Noise of the fan which is an indication of energy loss should be minimized in processing work;

4. Because different fans can fulfil the requirements for a specific job, the one to select will depend on the amount of use and the economic cost of power.

Table 12.2   Variations in Similarity Relations in Fans

Variations of V_s, H, P and HP			
With D, n, ϱ	**With ϱ**	**With D_2**	**With n**
$V_{s1} = V_{s2} \dfrac{D_1^3 \, n_1 \, \eta_{vol1}}{D_2^3 \, n_2 \, \eta_{vol2}}$	$V_{s1} = V_{s2}$	$V_{s1} = V_{s2} \dfrac{D_1^3}{D_2^2}$	$V_{s1} = V_{s2} \dfrac{n_1}{n_2}$
$H_1 = H_2 \dfrac{D_1^2 n_1^2 \eta_{h1}}{D_2^2 n_2^2 \eta_{h2}}$	$H_1 = H_2$	$H_1 = H_2 \left(\dfrac{D_1}{D_2}\right)^2$	$H_1 = H_2 \left(\dfrac{n_1}{n_2}\right)^2$
$P_1 = P_2 \dfrac{\varrho_1}{\varrho_2} \left(\dfrac{D_1}{D_2}\right)^2 \left(\dfrac{n_1}{n_2}\right)^2 \dfrac{\eta_{h1}}{\eta_{h2}}$	$P_1 = P_2 \dfrac{\varrho_1}{\varrho_2}$	$P_1 = P_2 \left(\dfrac{D_1}{D_2}\right)^2$	$P_1 = P_2 \left(\dfrac{n_1}{n_2}\right)^2$
$HP_1 = HP_2 = \dfrac{\varrho_1}{\varrho_2} \left(\dfrac{D_1}{D_2}\right)^5 \left(\dfrac{n_1}{n_2}\right)^3 \dfrac{\eta_2}{\eta_1}$	$HP_1 = HP_2 \dfrac{\varrho_1}{\varrho_2}$	$HP_1 = HP_2 \left(\dfrac{D_1}{D_2}\right)^5$	$HP_1 = HP_2 \left(\dfrac{n_1}{n_2}\right)^3$

When selection is guided by maximum capacity, V_s, and the head H required to move a given mass of air, it is noted that possible errors in calculating head loss exist. Therefore, fans are selected which are rated for a capacity of $1.05 V_s$ and a head of $1.1 H$. For forced-draft fans and flue-gas exhausters, fans of higher capacity are selected ($1.10 V_s$ and $1.2 H$). In catalogues, the data presented refer to standard conditions of $T = 293 K$ and $H_0 = 760$ mm Hg. So, select from the catalogue.

$$V_{s\,std} = (1.05 - 1.10)V_s \quad and \quad H_{std} = (1.10 - 1.20)(\varrho_{std}/\varrho)$$

where

$$\varrho_{std} = air\ density\ at\ NTP$$

12.2.10   Differences between axial and centrifugal fans (Gamble, 1996)

From the above, there are differences between axial and centrifugal fans which are summarized as follows:

- Axial fans can maintain higher efficiencies at various loads than can constant-speed centrifugal fans controlled by inlet vanes.
- Higher capital cost of axial fans over centrifugal fans frequently can be offset by operating cost savings over the life of the fans.
- Centrifugal fans consist of only one moving part, the fan rotor weldment; axial fans contain many moving parts.
- Adjustable pitch axial fans are more complex than centrifugal fans and require more frequent maintenance than centrifugal fans of equivalent capacity and must be maintained at a higher level of balance due to lower tolerance for operating vibration.
- Axial fans are more sensitive than centrifugal fans, when used to induce draft, to performance degradation caused by fan blade corrosion and trust bearing failure.

- To maintain variable and controlled pressure or flow conditions, the fans have capacity regulators, inlet or outlet dampers, two-speed or variable speed controls for centrifugal fans and adjustable pitch blade control for axial fans.

Axial fans are much more prone to operation in a stall condition (an aerodynamic phenomenon that occurs when fan operation is attempted at a condition that is beyond the fan pressure capacity) than are centrifugal fans. In such a condition, flow separation occurs around the blade, the fan becomes unstable, accompanied by noise and vibration, and may lead to fatigue failure of the rotating blades. With properly sized axial fans operating within the normal system resistance, probability of stall is low. Over-sized fans or increased system resistance (through damper closure, system plug, etc.) increases the probability to stall.

12.3 Blowers

Machines operating at gas pressure ratios $\varepsilon > 1.15$, yet without artificial cooling, are referred to as blowers. Blowers, which are essentially fans, are part of the principal components of air flow systems of farm machinery. They are an integral part of smaller furnace-blower systems for residential heating or drying or the impeller-blower system for elevating grains or chopped forage vertically into silos, in sprayers for pesticides, in combined harvesters for grains and cotton, even in threshers, cleaners and driers etc.

The air-flow characteristics are usually for the system and not the blower alone. The system includes the duct, circular or spiral casing, the filter and the blower. Blowers are usually driven by belt but are also being driven directly from the tractor PTO. The speed is limited to 540 rpm with most tractors. A greater diameter is required to secure the same peripheral speed as the fast-turning belt-driven impellers.

Centrifugal blowers usually used in farm machinery have few blades (Z = 2 to 6) and deliver air at either low pressure. ($H N/m^2 < 1000$). or medium pressure ($1000 N/m^2 < H < 3000 N/m^2$). The pneumatic transport equipment used to convey hay, grain and straw products is sometimes further equipped with a high-pressure blower (H >3000 N/m^2). The grain is primarily thrown rather than blown. The grain is introduced near the centre of the impeller rotor and leaves the blade tips at their peripheral speeds higher than the airflow velocity. The air provides some resistance to the material upward movement. The material slows down due to the effects of the air drag gravity and the friction with the duct or pipe walls. The blower supplies enough air to impart enough kinetic energy to help carry the material to the desired height after it loses its initial velocity.

The impeller is the principal component of a blower. The impeller may be made of straight or curved blades as shown in Fig. 12.12. The straight ones may be radial (Fig.12.12a) or curving backwards (Fig, 12.12b) and are used in farm machinery for providing flow rates at low medium pressures. The curved blades may be bent backwards (Fig 12.12c), bent backwards with their tips in the radial direction (Fig. 12.12d), or bent forward (Fig.12. 12e). The curved blades bent forward develop higher pressures than the others.

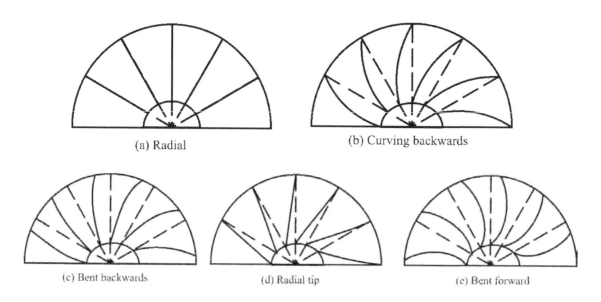

(a) Radial   (b) Curving backwards

(c) Bent backwards   (d) Radial tip   (e) Bent forward

Fig. 12.12: Straight and Curve blades of the impeller

12.3.1 Operating Parameters of Blowers

The velocity vectors of the impeller of a blower are given in Fig 12.13.
The curved vane 1-2 has radius r_1 at entry and r_2 at the exit of the impeller. If the vane imparts a certain absolute velocity V_a to the air at some angle ß to the radius vector, r (OA), the theoretical pressure of the air developed by the blower (ignoring resistance) is given in Eqn. 12.48.

$$H_t = N_t/V_{sa} = \varrho_a(U_2V_{o2} - U_1V_{o1}) \qquad N/m^2 \qquad \qquad \dots 12.48$$

The Euler equation has N_t = energy absorbed by the air due to blower action (Nm/sec.), V_{sa} = volume flow rate of air (m³/sec), ϱ_a = density of air (kg/m³),

U_1, U_2 = peripheral velocities at inlet and outlet of vanes (m/sec), V_{o1}, V_{o2} = projected velocities perpendicular to radius vector at inlet and outlet points (m/sec.).
The velocities are got using Eqns. 12.49, 12.50 and 12.51.

At point A, $\quad V_o = U - V_t \, Sin \, \alpha = (1 - [V_t \, Sin \, \alpha/U]) \qquad \dots 12.49$

$\therefore$ *In* ΔABC:

$$V_t/U = Sin \, (90 - \beta)/Sin \, (\alpha + \beta) = Cos \, \beta/Sin \, (\alpha + \beta) \qquad \dots 12.50$$

$$\therefore \quad V_o = U(1 - [Sin \, \alpha Cos \, \beta/Sin(\alpha + \beta)]) = U(Cos \, \alpha Sin \, \beta/Sin \, (\alpha + \beta)) \qquad \dots 12.51$$

$\qquad Also \; U = \omega r \quad and \quad \omega = \pi n/30$

Substituting these values in the Euler Eqn.12.48 gives Eqn. 12.52.

252

$H_t/\varrho_a n^2 r_2^2 = (\pi^2/900)[(Cos\,\alpha_2 Sin\,\beta_2/Sin(\alpha_2 + \beta_2)) -$
$(r_1/r_2)^2(Cos\,\alpha_1 Sin\beta_1/Sin(\alpha_1 + \beta_1))] = A$ ...12.52

For continuity, we have in Eqn. 12.53

$Q_a = 2\pi r_2 B V_{r2}$ ...12.53

where

V_{r2} = component of absolute velocity of air at the exit from the vane in the direction of the radius; B = width of the vane.

In Fig. 12.13, we get V_{r2} from Eqn. 12.54.

$V_{r2} = V_{o2}/Tan\,\beta_2 = U_2 Cos\,\alpha_2 Sin\beta_2/Tan\,\beta_2 Sin(\alpha_2 + \beta_2)$

$= U_2 Cos\,\alpha_2 Cos\,\beta_2/Sin\,(\alpha_2 + \beta_2) = \pi n r_2 Cos\,\alpha_2 Cos\,\beta_2/30 Sin(\alpha_2 + \beta_2)$
                    ...12.54

When substituted in Eqn.12.53, we obtain Eqn. 12.55.

$Q_a/Br_2^2 n = 2\pi^2 Cos\,\alpha_2 Cos\,\beta_2/30 Sin(\alpha_2 + \beta_2) = K$ ...12.55

Since the selection of the blowers is based on the principle of mechanical similarity, the linear dimensions (r_1/r_2; B_1/B_2) are proportional and the angular dimensions (α_1, α_2, β_1 and β_2) are equal, then Eqn. 12.52 and Eqn.12.55 are constants. Thus, it could be seen that: -

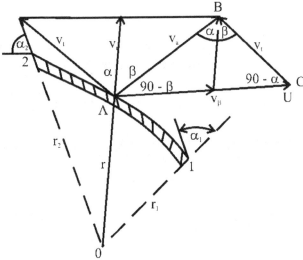

Fig. 12.13: Velocity vectors of the air at the impeller

(i) The theoretical pressure head developed by the blower varies proportionally to the square of the speed of rotation of the impeller (rpm) given as Eqn. 12.56.

$$H_t = A\varrho_a r_2^2 n^2 \qquad \qquad \dots 12.56$$

(ii) The flow rate of air is proportional to the speed (rpm) of the impeller given as Eqn. 12.57.

$$Q_a = KBr_2^2 n \qquad \qquad \dots 12.57$$

(iii) The energy absorbed by the air varies proportionally to the cube of the speed of rotation of the impeller (rpm) given as Eqn. 12.58.

$$N_t = H_t Q_a = AKB\varrho_a r_2^4 n^3 = CB\varrho_a r_2^4 n^3 \qquad \qquad \dots 12.58$$

where
$C = AK$. = constant.

12.3.2 Blower Characteristics

The quantitative characteristics of blowers may be dimensional and non-dimensional: -

(i)Dimensional characteristics are graphic relationships set up between V_{sa} = volume flow rate of air, H = total pressure head, H_{st} = static head, H_d = dynamic head, N = input power, η_{st} = static efficiency and η = total efficiency. They are plotted by measuring the total and dynamic head at different flow rates (Eqn. 12.59) where; -

$$Q_a = \varrho_a V_a A_o \qquad \qquad \dots 12.59$$

where
V_a = velocity of airstream at exit of the blower (m/s); A_o = cross-sectional area of exit (m^2); Q_a =mass flow rate of air (kg/s).
$V_a = 1.29(H_d)^{\frac{1}{2}} = (2H_d/\varrho_a)^{\frac{1}{2}}$ m/s
ϱ_a = density of air at exit of the blower = 1.2 kg/m^3 at 20^{0}C and 10.3 x 1.0^4 N/m^2 atmospheric.
A_o = area of cross-section of exit of the blower.

The **total and static efficiencies** are given in Eqns. 12.60a and 12.60b.

$$Total\ efficiency\ \ \eta = 1.0^{-3}\ (Q_a/\varrho_a)\ (H/N) \qquad \dots 12.60a$$

$$Static\ efficiency\ \ \eta_{st} = 1.0^{-3}\ (Q_a/\varrho_a)\ (H_{st}/N) \qquad \dots 12.60b$$

where
Q_a = mass flow rate of air, kg/sec; ϱ_a = kg/m^3; H = N/m^2 and N = kW

(ii) Non-dimensional characteristics are derived from dimensional characteristics by referring the parameters to the impeller speed of 1000 rpm and plotting them. Also, the index K_a, which characterizes the resistance of the screen, i.e. the friction losses due to flow over duct walls due to change in flow direction etc. are also plotted. K_a is given by Eqn. 12.61.

$$K_a = (H_d/(H_{st} + H_d))^{1/2} \qquad \qquad ...12.61$$

These characteristics determine the properties of a series of similar blowers.

12.3.3 Selection of Blowers

Even though it has been said that the selection of blowers is based on the principle of mechanical similarity, i.e. having the linear dimensions proportional and angular dimensions equal, there are principal data for the selection of blowers. These are: -

(i) V_a = velocity at the exit of duct
(ii) Q_a = requisite flow rate delivered by the duct
(iii) H = total pressure the blower must develop

The velocity V_a (Eqn. 12.62) must exceed the critical velocity (V_{cr}) as in the pneumatic transport. However, the excess velocity coefficient α_a differs according to the material being handled.

$$V_a = \alpha_a V_{cr} \qquad \qquad ...12.62$$

where
α_a = 1.1 - 1.7 (straw); 1.9 - 3.7(chaff); 2.5 - 5.0 (husk); 1.5 - 3.0 (broken ears of corn).
The flow rate Q_a is selected based on the permissible number of impurities per unit mass of air (Eqn. 12.63),

$$Q_a = q_l/\lambda \qquad \qquad ...12.63$$

where λ = concentration coefficient. For blowers through which impurities are not transported like combines and winnowers $\lambda = 0.2 - 0.3$; for dust handling blowers $\lambda = 0.14 - 0.15$.

The head H developed by the blower (Eqn. 12.64) must be: -

$$H = H_{st} + H_d \qquad \qquad ...12.64$$

where H_{st} consists of all the losses in the blower, friction losses over the walls of duct, losses due to direction change, losses over sieve and gates within the ducts, resistance of sieve and the pile of grains on it. The dynamic head (H_d) is given as in Eqn. 12.65.

$$H_d = dynamic\ head = \varrho_a V_a^2/2 \quad N/m^2 \qquad \qquad ...12.65$$

Because the calculations of H_{st} and H are difficult, one must make assumptions, during design calculations. While selecting blowers, the various resistance is estimated by analogy with existing systems. Given the coefficient K_a, H_{st} is obtained using Eqn.12.61 and knowing actual pressure H, the theoretical head is got from Eqn. 12.66.

$$H_t = H/\eta \qquad\qquad \text{...12.66}$$

where

η is got from Eq.12.60a.

On the basis of the above quantities and using Eqn.12.52, Eqn.12.55 and Eqn.12.58, we obtain the following parameters for the blowers: -

(i) r_2 (ii) n   and   (iii) N = power input.

12.3.4 Impeller Blowers

The impeller blower is a simple machine useful for elevating grain vertically. Tests indicate that wheat (at $12\frac{1}{2}$ % m.c.); corn (at 10 % m.c.) and rice (at $8\frac{1}{2}$ % m.c.) are not appreciably damaged as to cracking and germination at impeller peripheral speeds below 21 m/sec. Power consumption is at a minimum when the peripheral speed is only enough to elevate the grain to the desired height.

The actual delivery height falls short of the theoretical height to an increasing extent as the internal friction increases with higher rates of grain flow (Fig.12.14). It is observed that for wheat, higher tip velocity is required to deliver greater rates to a given height and that 6.35 t/h can be lifted only 6 m without exceeding 21 m/sec (Richey, et al., 1961).

For impeller blowers, the radial paddles give best results. Grain damage usually occurs at the initial impact with the blade, and it is important that the grain be introduced at the inner circle of the blades where the velocity is the lowest. It was found possible to increase the allowable peripheral speed to 30 m/sec when the paddles were covered with rubber to reduce cracking.

A **forage blower** used to elevate chopped forage into the silo or barn is essentially an impeller blower, which throw the material up the delivery pipe while supplying enough air to carry it through to a destination. They rely on passing forage through the impeller/fan in order to provide the initial momentum. The conditions of pneumatic conveying are maintained: -
(i)   Adequate airspeed and ratio between the amount of air and the amount of crop handled;
(ii)   Avoidance of roughness or bends which tends to cause the material to be caught up and lead to blockages.

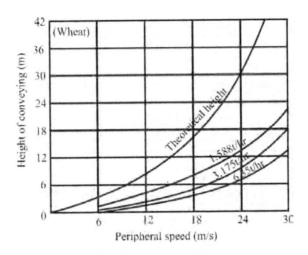

Fig. 12.14: The height of conveying difference width capacities at different peripheral speeds

However, the building or container into which the forages are "blown" needs to be designed in such a way that easy escape of the exhausted air is provided for. For high-capacity blower propelling forage into a sealed tower silo, both conveying and distribution are inefficient unless an adequate air escape, preferably some distance from the inlet, is provided. Here, recommended peripheral velocity ranges from 35 m/sec to 50 m/sec.

The kinetic energy amounts to 0.26 kW-h/t for 45 m/sec. Since only 0.03 kW-h is theoretically required to elevate the forages 12 m, it is apparent that blowers are inefficient as much energy is lost by friction of the material on the paddle blades and on the inside of the housing.

In Fig 12.15, the maximum wing length L, which can be unloaded is given by Eqn. 12.67.

$$L = R(1 - [1/Cosh\ \theta]) \qquad \qquad ...12.67$$

where
R = radius to wing tip (cm) and θ = outlet angle (rad).

The speed of rotation is not a factor here. But the formula is of value in determining the angular rotation required for a blade to unload a given amount of material and the corresponding size of the discharge opening required. The direction and velocity of discharge, shown in Fig.12.16 is given by Eqn. 12.68.

$$Tan\ \psi = Sinh\ \theta / Cosh\ \theta \qquad \qquad ...12.68$$

where ψ = angle of discharge from the tangential and θ = angle of rotation past the beginning of discharge opening i.e the start of particle's radial motion in radians, and the discharege velocity is given by Eqn. 12.69.

257

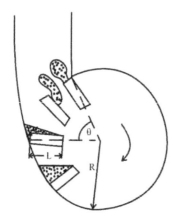

Fig. 12.15: Maximum wing length that can be unloaded

$$V_D = V_T / Cos\ \psi \qquad\qquad ...12.69$$

where

V_D = discharge velocity and V_T = tangential velocity.

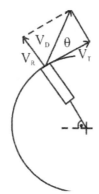

Fig. 12.16: Direction and velocity of discharge of an impeller blower

In order to achieve minimum housing friction, the feed opening (duct) should be located so that the material slides to the end of the blades just as the discharge opening is reached. A typical impeller blower has a six-blade 53 cm diameter rotor with 15 – 23 cm diameter delivering pipe and is rated at 16 – 36 m³/hr. However, hay at 35 - 45% moisture content is the material most liable to clog the duct of the machine.

However, for **forage blowers**, Totten and Millier (1966) suggested that their total power requirements may be computed using Eqn. 12.70.

$$HP = Qr^2\omega^2/2\eta \qquad\qquad ...12.70$$

where

Q = Mass flow rate (kg/s); η = Power efficiency of forage blower; r = Paddle tip radius (m); ω = Angular velocity (rad/s).

258

While the blower design factors include the blower size, no. of blades, angle of paddle-slant, clearance between paddle tip and the scroll, the location and size of the material inlet opening and the size (diameter and length) of the conveying pipe; for the satisfactory operation of the blower, the conveying height, blower speed, bulk density and coefficient of friction of the material being conveyed should be included. These affect the efficiency of the blower which increases with paddle speed, paddle loading and reduced friction (Pettingill and Millier, 1968) as shown in Table 12.3.

Table 12.3: **Blower efficiency (%) for various speeds, feed rates, design changes and air inlet conditions (Pettingill and Millier, 1968)**

Feed rate (kg/min)	Blower speed (rpm)	Air door conditions	
		Open	Closed
408	600	24.5	25.7
544	800	28.4	30.7
680	600	24.6	25.9
907	600	25.5	26.9
Design conditions at 600 rpm and 680 kg/min			
Normal system		24.60	25.90
Inlet cut-off		24.63	25.22
Teflon paddles		24.08	24.17
Teflon paddles and housing		24.70	26.01

12.4 SOLVED PROBLEMS

Q 12.1 A short capacity fan operating at 2900 rpm against 76 mm static pressures delivers 3200 m^3/s of air and requires 3.0 kW from a 4.5 kW motor. The fan owner does not want to change the motor. What are the system's maximum capacity, allowable speed increase and the static pressure under these conditions?

Data: $n_1 = 2900 \, rpm$; $P_1 = 76 \, mm$; $Q_1 = 3200 \, m^3/s$; $HP_1 = 3.0 \, kW$; $HP_2 = 4.5 \, kW$; $n_2 = ?$ $P_2 = ?$ $Q_2 = ?$

Solution: $HP_1/HP_2 = (n_1/n_2)^3$ $\therefore n_2 = n_1(HP_1/HP_2)^{1/3} = 2900(4.5/3.0)^{1/3} = \textbf{3320} \textbf{\textit{rpm}}$

$Q_1/Q_2 = n_1/n_2$ $\therefore Q_2 = Q_1 n_2/n_1 = 3200(3320/2900) = \textbf{3663} \, \textbf{\textit{m}}^3/\textbf{\textit{s}}$

$P_1/P_2 = (n_1/n_2)^2$ $\therefore P_2 = P_1(n_2/n_1)^2 = 76(3320/2900)^2 = \textbf{99.61} \textbf{\textit{mm}}$

Q 12.2 A fan with $D_1 = 0.8 \, m$ of $HP_1 = 3.0 \, kW$ operates at 2000 rpm carrying 200 m^3/s of air of $\varrho_1 = 1.30 \, kg/m^3$ with $\eta_1 = 0.85$ against $P_1 = 0.80 \, kN/m^2$. Design a similar fan to deliver $Q_2 = 275 \, m^3/s$ of air of $\varrho_2 = 1.12 \, kg/m^3$ with $\eta_2 = 0.95$. Find the motor HP_2 if the power reserve factor $m = 1.20$, and $\eta_{tr} = 0.90$. Assume $\eta_v = \eta_h = 100\%$. What are the maximum Q and H of the designed fan?

Data: $Q_1 = 200 \, m^3/s$; $Q_2 = 275 \, m^3/s$; $\varrho_1 = 1.30 \, kg/m^3$; $\varrho_2 = 1.12 \, kg/m^3$; $\eta_1 = 0.85$; $\eta_2 = 0.95$; $P_1 = 0.80 \, kN/m^2$; $HP_1 = 3.0 \, kW$; $n_1 = 2000 \, rpm$; $m = 1.20$; $\eta_{tr} = 0.90$; $\eta_v = \eta_h = 100\%$; $D_1 = 0.8 \, m$

Solution: $Q_1/Q_2 = n_1/n_2$ $\therefore n_2 = n_1(Q_2/Q_1) = 2000(275/200) = \textbf{2750} \, \textbf{\textit{rpm}}$

$Q_1/Q_2 = (D_1/D_2)^3$ $\therefore D_2 = \sqrt[3]{(Q_2 D_1^3/Q_1)} = \sqrt[3]{(275 x 0.8^3/200)} = \textbf{0.89} \, \textbf{\textit{m}}$

$P_1/P_2 = (\varrho_1/\varrho_2)(D_1/D_2)^2(n_1/n_2)^2$ $\therefore P_2 = 0.8(1.12 x 0.89^2 x 2750^2/1.30 x 0.80^2 x 2000^2)$

$= \textbf{1.61} \, \textbf{\textit{kN}}/\textbf{\textit{m}}^2$

$HP_1/HP_2 = (\varrho_1/\varrho_2)(D_1/D_2)^5(n_1/n_2)^3(\eta_1/\eta_2)$

$\therefore HP_2 = 3.0(1.12 x 0.85^5 x 2750^3 x 0.95/1.30 x 0.80^5 x 2000^3 x 0.85) = \textbf{6.82} \, \textbf{\textit{kW}}$

$HP_{motor} = mQgH\varrho/\eta_{tr}\eta_1\eta_2 = mQP_2/\eta_{tr}\eta_1\eta_2 = 1.2 x 275 x 1.61/0.9 x 0.85 x 0.95 = \textbf{7.31} \, \textbf{\textit{kW}}$

$H_2 = P_2/\varrho_2 g = 1.16 x 10^3/1.3 x 9.81 = 126.24m$ $\therefore H_{max} = 1.1 H_2 = \textbf{138.87} \, \textbf{\textit{m}}$

$Q_{max} = 1.05 Q_2 = 1.05 x 275 = \textbf{288.75} \, \textbf{\textit{m}}^3/\textbf{\textit{s}}$

Q 12.3 Corn is dried by a fan drawing 9500 m³/s of hot air of density 0.90 kg/m³ from an oven against a pressure of 80 mm. The fan operates at 1000 rpm and requires 9.5 kW. Assuming the air loses its heat and now has density 1.2 kg/m³, what happens to the pressure and the power? What do you observe about the selection of power in this circumstance?

Data: $\varrho_1 = 0.90 \, kg/m^3$; $\varrho_2 = 1.2 \, kg/m^3$; $Q_1 = Q_2 = 9500 \, m^3/s$; $P_1 = 80 \, mm$; $n_1 = 1000 \, rpm$; $HP_1 = 9.5 \, kW$.

Solution: $P_1/P_2 = \varrho_1/\varrho_2 = HP_1/HP_2$ ∴ $P_2 = P_1 \varrho_2/\varrho_1 = 80 \, x \, 1.2/0.9 = \mathbf{106.7 \, mm}$

$HP_2 = HP_1 \varrho_2/\varrho_1 = 9.5 \, x \, 1.2/0.9 = \mathbf{12.7 \, kW}$

Power is selected based on maximum density at lower temperature.

Q 12.4 An axial fan moving 3.2 m³/s of air at 1500 rpm has a hub diameter ratio of 0.8 and an absolute to tangential velocity ratio of 0.55. Find the fan parameters.

Data: *Hub diameter ratio* $= v = D_h/D_{out} = 0.8$; $V_s = 3.2 \, m^3/s$; $n = 1500 \, rpm$
Absolute/tangential velocity ratio $= k_\varphi = 0.55$.

Solution: $D_{out} = 2.9 \, x \, \sqrt[3]{[V_s/v(1 - v^2)(k_\varphi n)]}$

$= 2.9 \, x \, \sqrt[3]{[3.2/0.8(1 - 0.8^2)(0.55x1500)]} = \mathbf{0.69 \, m}$

D_h = hub diameter = 0.69 x 0.8 = **0.55 m**

Vane length $= l_v = (D_{out} - D_h)/2 = (0.69 - 0.55)/2 = \mathbf{0.07 \, m}$

$D_{rms} = \sqrt{((D_{out}^2 + D_h^2)/2)} = \sqrt{((0.69^2 + 0.55^2)/2)} = \mathbf{0.624 \, m}$

D$_{rms}$ = root mean square diameter

$C_a = V_s/[0.785D_{out}^2(1 - v^2)] = 3.2/[0.785(0.69)^2(1 - 0.8^2)] = \mathbf{23.8 \, m/s}$

$or \; C_a = \pi k_\varphi v D_{out} n/60 = \pi x 0.55 x 0.8 x 0.69 x 1500/60 = \mathbf{23.8 \, m/s}$

$Since \; C_a = k_\varphi \mu$ ∴ $\mu = C_a/k_\varphi = 23.8/0.55 = \mathbf{43.3 \, m/s}$

Q 12.5 A manufacturer wishes to project data obtained for a $D_1 = 30 \, cm$ diameter fan to a $D_2 = 60 \, cm$ diameter fan. At one operating point, the 30 cm fan delivers 8000 m³/min of 27°C air against 76 mm of static pressure. This requires 700 rpm and 2.0 kW motor. What will be the projected equivalent parameters for the 60 cm fan at the same rpm?.

Data: $D_1 = 30 \, cm$; $D_2 = 60 \, cm$; $V_{s1} = 8000 \, m^3/min$; $P_1 = 76 \, mm$;
$n_1 = n_2 = 700 \, rpm$; $HP_1 = 2.0 \, kW$; $T_{s1} \, speed = \pi Dn \, cm/min$.

Solution:

$$V_{s2} = V_{s1}(D_2/D_1)^3 = 8000(60/30)^3 = \textbf{64000 } \boldsymbol{m^3/min}$$

$$P_2 = P_1(D_2/D_1)^2 = 76(60/30)^2 = \textbf{304 } \boldsymbol{mm}$$

$$HP_2 = HP_1(D_2/D_1)^5 = 2(60/30)^5 = \textbf{64 } \boldsymbol{kW}$$

$$T_{s1} = \pi x30x700 = \textbf{65973.4 } \boldsymbol{cm/min}$$

$$T_{s2} = T_{s1}(D_2/D_1) = 65973.4(60/30) = \textbf{131,946.8 } \boldsymbol{cm/min}$$

Q 12.6 An existing $3.0\ kW$ fan of hub diameter $12\ cm$ exhausts a gas of density $1.25\ kg/m^3$ at a speed of $1800\ rpm$ with a mechanical efficiency of 0.85. To exhaust another gas of density $0.95\ kg/m^3$ with an efficiency of 0.95, what will be the horse power of a fan of hub diameter $15\ cm$ that will do the job at $1500\ rpm$?

Data: $HP_1 = 3.0\ kW$; $D_1 = 12\ cm$; $\varrho_1 = 1.25\ kg/m^3$; $n_1 = 1800\ rpm$; $\eta_1 = 0.85$; $\eta_2 = 0.95$; $\varrho_2 = 0.95\ kg/m^3$; $D_2 = 15\ cm$; $n_2 = 1500\ rpm$; $HP_2 = ?$

Solution: $HP_2 = HP_1 . (\varrho_2\eta_1/\varrho_1\eta_2) . (D_2/D_1)^5 (n_2/n_1)^3$

$$= 3.0(0.95/1.25).(0.85/0.95).(15/12)^5.(1500/1200)^3 = \textbf{3.60 } \boldsymbol{kW}$$

CHAPTER 12 Review Questions

12.1 Differentiate between fans, blowers and compressors and state how, why and where they are used in agricultural processing.

12.2 Describe briefly the configuration of fans and the types of fans. What are the functions of fans?

12.3 What are the characteristics of the different types of centrifugal fan blades?

12.4 Explain in brief the parameters to be considered when designing axial-flow fans giving the accompanying equations. How is axial fan flow output controlled?

12.5 What is fan capacity regulation? What fan parameters does it vary; and how would you adjust fan capacity?

12.6 Give the expression for the specific energy imparted to the air and what does the added energy do to the flow?

12.7 Describe system resistance and give the equations of the two losses that make it up.

12.8 Define total efficiency, static efficiency, hydraulic efficiency, volumetric efficiency, mechanical efficiency as in centrifugal fans. What is their relationship?

12.9 State the six (6) fan affinity laws. Give the equations for Fan Laws A, B and C.

12.10 State the factors that affect fan selection. Give the differences between axial and centrifugal fans.

12.11 Describe fan performance characteristics.

12.12 What does the principle of mechanical similarity for blower selection mean and what three conclusions can be drawn therefrom?

12.13 Describe a blower.

12.14 What data do you need to select a blower? Give the corresponding equations and the parameters to be obtained for the blower.

12.15 What are the functions of impeller and forage blowers? What are the components of a blower?

12.16 A fan of 2.0 kW with outer diameter 0.85 m operates at 1500 rpm carrying 4.0 m^3/s of air of $\varrho_a = 1.20 \, kg/m^3$; with 85% efficiency against a total pressure of 0.4 kN/m^2. Design a similar fan to deliver 5.0 m^3/s of air of $\varrho_a = 1.10 \, kg/m^3$ with efficiency 95%. Determine the motor HP if power reserve factor = 1.2, transmission efficiency = 90%. Assume $\eta_v = \eta_h = 100\%$. What will be the maximum capacity and the head of the fan you will select? *(n_b = 1875 rpm; D_b = 0.92 m; P_b = 0.671 kN/m²; HP_b = 4.76 kW; HP_m = 5.53 kW; H_b = 62.2 m, H_{max} = 68.4 m, V_{smax} = 5.25 m³/s)*

12.17 If an axial flow fan moves 2.40 m^3/s of air at 1440 rpm, and the hub diameter ratio is 0.75 and the absolute to the tangential velocity ratio is 0.60, find the hub diameter, vane length, root-mean-square diameter and peripheral velocity. *(D_{out} = 0.592 m; D_n = 0.443; L_v = 74 mm; D_{rms} = 0.522 m, C_a = 20.0 m/s; μ_h = 33.3 m/s)*

12.18 An existing 2.5 kW fan of D_h = 10 cm exhausts gas of $\varrho_g = 1.20 \, kg/m^3$ at n_a = 1750 rpm with $\eta_m = 0.85$. To exhaust another gas of $\varrho_g = 0.90 \, kg/m^3$ with $\eta_m = 0.95$, what will be the HP of the fan of hub diameter = 12 cm that will do the job at 1500 rpm? Determine the capacity and pressure ratios of the fans for $\eta_v = \eta_h = 100\%$. *(HP = 2.63 kW ; Q_a/Q_b = 1.48; P_a/P_b = 0.79)*

12.19 A short capacity fan operating at 2700 rpm against 76 mm static pressure delivers 3600 m^3/s of air and requires 2.84 kW from a 5.0 kW motor. The owner of this fan does not want to spend any money to change the motor. What is the maximum capacity from this system with the existing 5 kW motor? What is the allowable speed increase? What is the static pressure under the new conditions? *(n_b = 3260 rpm; Q_b = 4347 m³/s; P_2 = 110.8 mm).*

12.20 To dry grain, a fan draws 18000 m^3/s of air from an oven at $116°C$ against a pressure of 63.5 mm. The fan operates at 800 rpm and requires 10 kW. Assume the oven loses its heat and the air is at $21°C$. What happens to the pressure and the power required? What do you observe about the selection of power? Assume $\varrho_{21°C}$ = 1.2015 kg/m^3; $\varrho_{116°C}$ = 0.8971 kg/m^3. *(P_2 = 85 mm; HP_2 = 13.4 kW; Always select power based on max. density at lower air temperature).*

12.21 A pneumatic system is used to exhaust air of 0.9 kg/m^3 into a 5.4 m high silo at 6.6 m/s and 0.95 kN/m^2. At the 2.5 m^2 area outlet, the air moves at 3.8 m/s and 1.26 kN/m^2 with a head loss of 0.75 m. Find the total and theoretical heads developed by the exhausters, the hydraulic efficiency and the specific energy of transfer. *(H_t = 32.5 m; H_{th} = 33.25 m; η_n = 0.98; L_{th} = 326.1 J/kg)*

12.22 In the exhausters in Q. 17, what power is developed if η_m = 0.90 and η_v = 0.92? What extra amount of air is introduced into the system? Estimate the power of the fan motor if it is directly mounted with a power reserve factor = 1.15. Assume $\varrho_g = \varrho_{air}$. *($\eta$ = 0.81; HP = 3.4 kW; ΔQ = 0.76 m²/s; Q_1 = 8.74 m³/s; HP_m = 3.87 kW)*

12.23 The material flow rate of a forage blower is 10.56 kg/s when impeller speed is ω = 500 rpm. If η = 25%, what will be the required power? Assume paddle tip radius to be 0.5 m *(HP = 14.48 kW)*

REFERENCES

Abou Taleb, F. S. and J. Abdallah. 2017. Experimental study of an air lift pump. Engineering Technology and Applied Science Research. 7(3):1676-1680.

Ahmadi, S., S. Moosazadeh, M. Hajihassani, H. Moomivand and M. M. Rajaei. 2019. Reliability, availability and maintainability analysis of the conveyor system in mechanized tunnelling. Measurement. 145: 756-764

Alexandrov, M. P. 1981. Materials Handling Equipment. Mir Publishers, Moscow.

Allegri, Theodore H., Sr. 1992. Materials Handling Principles and Practice. Malabar, Florida: Krieger Publishing Company.

Altamuro, M. V. and V. H. Hawkins. 1986. Materials Handling. In. E. A. Avallone and T. Baumeister III (eds). Marks Standard Handbook for Mechanical Engineers, 9th Edition. McGraw-Hill Book Company, New York.

American Chain Association (ACA), Second Edition, "Standard Hand Book of Chains- Chains For Power Transmission and Material Handling," 2006.

Anon 1954. Hot dust is conveyed pneumatically from preceptors to furnaces. Engineering and Mining Journal; July 1954. P. 91.

ASAE Standards. 2005. Data D241.2

ASHRAE 1972. American Society of Heating, Refrigeration and Air-conditioning engineering Handbook of Fundamentals. New York. ASHRAE

Asoegwu, S. N. 1988. Estimation of leaf area of two okro (*Abelmoschus esculentus*) varieties through leaf characteristics. Indian Journal of Agricultural Science. 58(11) 862 - 863.

Asoegwu, S. N. 1989. The Industrial Potentials of Some Nigeria's Fruits and Vegetables. Occasional Paper No 20. National Horticultural Research Institute (NIHORT), Ibadan 10pp.

Asoegwu, S.N. 1991. Yield response of late season plantain to frequential irrigation. The Nigerian Engineer. 26(2): 1 - 12

Asoegwu, S.N. 1995. Some physical properties and cracking energy of conophor nuts at different moisture content. International Agrophysics. 9:131 - 142

Asoegwu, S.N. 1996. Firmness and storability of plantains fruits under ambient temperatures. International Agrophysics. 10: 37 - 41.

Asoegwu, S.N., G.I. Nwandikon and O.P. Nwammuo. 1998. Some mechanical properties of plantain fruit. International Agrophysics. 12: 67 - 77.

Asoegwu, S.N. and A. I. Ogbonna. 1999. Corrosion susceptibility of some metals in a simulated egusi-melon mucilage extractor corrodent. Nigerian Corrosion Journal. 2: 85 93.

Asoegwu, S.N. and J.O. Maduike. 1999. Some physical properties and cracking energy of *Irvingea gabonensis* (Ogbono) nuts. Proceedings of the Nigerian Institution of Agricultural Engineers 17: 131 - 137.

Barre, H. J. 1958. Flow of bulk granular materials. Agricultural Engineering 39(9): 534 - 536, 539.

Bartl, G., M. Krystek, A. Nicolaus and W. Giardini. 2010. Interferometric determination of the topographies of absolute sphere radii using the sphere interferometer of PTB. Measurement Science and Technology, 21(11), 115101.

Baten, W. D. and R. E. Marshall. 1943. Some methods for approximate prediction of surface area of fruit. Journal of Agricultural Research 66(10): 357 - 373.

Berry, P. E. 1958. Research on oscillating conveyors. Journal of Agricultural Engineering Research 3(3): 249 - 259.

Besch, E. L., S. J. Sluka and A. H. Smith. 1968. Determination of surface area using profile recording. Poultry Science. March 1968.p.57

Beverly, G. L., A. W. Roberts and J. W. Hayes. 1983. Mechanics of high-speed elevator discharge. Bulk Solids Handling 3(4): 853 - 859.

Bhandari, V. B. 2003. Design of Machine Element, Tata McGraw Hill publishing company, eighth edition.

Boumans, G. 1985. Grain Handling and Storage (Development in Agricultural Engineering 4). Elsevier Science Publishers, Amsterdam. P. 122.

British Standard BS 4409. Part 3: 1982 (Also ISO 7119 - 7981). Screw Conveyors - Method for calculating drive power. British Standards Institution, London.

Bushell, E. and R. C. Maskell. 1960. Fluidized handling of alumina powder. Mechanical Handling 47(3): 126 - 131.

Butler, P. 1974. No-moving parts conveyor shifts dry powdered solids. Process Engineering. August 1974. p.65.

Carleton, A. J., J. E. P Miles and F. H. H.Valentin, 1969. A study of factors affecting the performance of screw conveyors and feeders. Trans ASME Journal of Engineering and Industry. (68-MH-22) 91(2): 329 - 334. https://doi.org/10.1115/1.3591565

CEMA Book No 350. Screw Conveyors 1971. Conveyor Equipment Manufacturers Association (USA).

CFIA 1978. Canadian Feed Manufacturing Technology Handbook. Canadian Feed Industry Association, Ottawa Ontario, Canada.

Chen, L.C. (2007). Automatic 3D surface reconstruction and sphericity measurement of micro spherical balls of miniaturized coordinate measuring probes. Measurement Science and Technology, 18, 1748-1755.

Cherkassky, V. M. 1985a. Jet Pumps. Chapter 17 of Pumps Fans Compressors. Mir Publishers. Moscow. P. 364 - 373.

Cherkassky, V. M. 1985b. Centrifugal Fans. Chapter 5 of Pumps Fans Compressors. Mir Publishers. Moscow. P. 174 - 198.

Chuma, Y., T. Shiga and M. Iwamoto. 1978. Mechanical properties of Satauma orange as related to design of containers for bulk transportation. Journal of Texture Studies. 9: 461 - 479.

Clark, N. N. and R. J. Dabolt. 1986. A general equation of air lift pump operating in slug flow. American Institution of Chemical Engineers Journal., 32: 56-64.

Colijn, H. and F. J. C. Rademacher. 1978. Vibratory Conveyors. Paper presented at the 11th Bulk Handling Seminar, Univ. of Pittsburgh (USA), December 7 - 8, 1978.

Colijn, H. 1985. Mechanical Conveyors for Bulk Solids (Studies in Mechanical Engineering 4). Elsevier Science Publisher, Amsterdam, The Netherlands.

Curray, J. K. 1951. Analysis of Sphericity and Roundness of Quartz Grains. M. S. Thesis in Mineralogy. The Pennsylvania State University, University Park, Pa.

Delwiche, M. J., S. Tang and J. J. Mehlschau. 1989. An impact force response fruit firmness sorter. Transactions of the ASAE. 32(1): 321 326.

Despotović, Z. and Stojiljkovic, Z. 2005. Current Controlled Transistor Power Converter for Driving Electromagnetic Vibratory Conveyor. Proceedings of the XIII International Symposium of the Power Electronics, N.Sad 2.XI-4.XI.2005, Vol.T1-2.1, pp 1-5 .

Despotović, Z., V. Šinik, S. Janković, D. Dobrilović and M. Bjelica. 2015. Some specific of vibratory conveyor drives. V International Conference Industrial Engineering and Environmental Protection 2015 (IIZS 2015) October 15-16th, 2015, Zrenjanin, Serbia.

Dixon, G. 1981. Pneumatic Conveying. In: Plastic Pneumatic Conveying and Bulk Storage, (ed) G. Butters. Applied Science Publishers.

Dong, M, and Q. Luo. 2011. Research and Application on Energy Saving of Port Belt Conveyor, Procedia Environmental Sciences, Vol. 10, Part A, P. 32-38, ISSN 1878-0296, https://doi.org/10.1016/j.proenv.2011.09.007.

Duckworth, R. A., L. Pullum, Addie, G. R. and C. F. Lockyear. 1986. The pipeline transport of coarse materials in a non-Newtonian carried fluid. Proceedings of Hydro Transport 10. Bermuda Human Resources Association (BHRA) Conference. Innsbruck. October 1986: p. 69 - 81.

Dumbangh, G. D. 1985. An analysis of drive methods for vibrating equipment used in bulk solids systems. Proceedings of 10th Powder and Bulk Solids Conference, Chicago, May 1985, p. 452 - 470.

FEM 2. 121.1985. Screw conveyors for bulk materials - recommendations for the design. Section II, Continuous Handling. September, 1985.

Frechette, R. J. and J. W. Zahradink. 1965. Surface area-weight relationships for McIntosh apples. Transactions of the ASAE 9(4): 526 - 527.

Gamble, K. A. 1996. Fans. In: Power Plant Engineering by Black & Veatch; Drbal, L. F. et al., (eds). Published by Kluwer Academic Publishers. Boston/Dorrecht/London. Chapter 10: 286 - 309.

Gleason, E., Schwenke, H. (1998). A spindless instrument for the roundness measurement of precision spheres. Precision Engineering, 22(1), 37-42.

Glugosh, D. and R. R. Knotek. 2019. Chain & Sprocket Systems and Maintenance. Power Transmission Engineering. June 2019: pp. 34-36. www.powertransmission.com

Goodyear Handbook of Belting. 1982. Goodyear Tire and Rubber Company, Akron, Ohio. USA.

Gould, Les. 1993. Application guidelines for accumulation conveyors. Modern Materials Handling. 48(5): 42-43.

Gunal, Ali K., E. Williams and S. Sadakane. 1996. Modeling of Chain Conveyors and Their Equipment Interfaces. Simulation Conference, 1996. Proceedings. Winter. Pp. 1107-1114

Gupta, D. and D. Dave. 2015. Study and performance of belt conveyor system with different type parameter. IJIRST - International Journal for Innovative Research in Science & Technology. 2(06): 2349-6010.

Gutman, I. 1968. Industrial Uses of Mechanical Vibrations. Business Books Ltd., London.

Hall, Carl W. 1958. Theoretical considerations in materials handling systems. Agricultural Engineering. 39(9): 524 - 529, 539.

Haneman, S. and H. K. Mocha. 1978. Vibration has wide range of practical uses. Bulk storage Movement Control, May/June 1978: p.101 -103.

Harris, W., K. Felton and G. Burkhardt. 1965. Design data for pneumatic conveying of chopped forage. Transactions of the ASAE 8: 194 - 195.

Harrison, A. and A. W. Roberts. 1983. Technical requirements for operating conveyor belts at high speed. Proceedings of the International Conference on Bulk Materials Storage Handling and Transportation. Newcastle, Australia, August 1983. p. 84 - 89.

Hawksley, P. G. W. 1951. The physics of particle size measurement. British Coal Utilization Research Association (BCURA) Bulletin. 15(4): 105 - 146.

Henderson, S. M. and R. L. Perry. 1955. Agricultural Process Engineering. John Wiley and Sons Ltd. New York.

Hightower, J. 1972. Hard Tomatoes Hard Times. Cambridge, MA: Schankman Publishing Co.

Hill, T. J. E. 1980. The application and design of vibratory conveyors. Proceedings Solidex 80 Conference. Harrogate U.K, March/April 1980 Paper B 1.

Hoshimov, A., A. Rustamov and J. Rozzokov. 2018. Industrial Conveyors' Taxonomy and Its Applications. Acta of Turin Polytechnic University in Tashkent: Vol. 8: Iss.3, Article 5.

Hudson, W. G. 1954. Why use pneumatic conveyors? Chemical Engineering, April 1954. pp 191 - 194.

Hughes, B. R. and J. T. A. Proctor. 1981. Estimation of leaflet, leaf and total leaf area of *Panax quinquefolius* L. using linear measurements. Journal of the American Society of Horticultural Science. 106/2): 167 - 170.

Janecki, D., S. Adamczak and K. Stępień. 2012. Problem of profile matching in sphericity measurements by the radial method. Metrology and Measurement Systems. XIX(4):703-714.

Kadam, A. J. and S. V. Deshpande, 2015. Design & development of conveyor chain outer link by using composite material. International Journal of Innovations in Engineering Research and Technology [IJIERT]. 2(12): 1-8.

Kate, A. 2018. Design of Material Handling Equipment Suitable for Agro Commodities. Advances in Agricultural Engineering Annual Technical Volume 2 pp. 54-69

Keck, H. and J. Goss. 1965. Determining aerodynamic drag and terminal velocities of seeds in free fall. Transactions of the ASAE 8: 553 -554.

King, B. C. 1980. The application and design of en-masse conveyors: Proceedings Solidex 80 Conference. Harrogate, U.K. March/April 1980, Paper A3.

Khanam, T., W.N. Syuhada Wan Ata and A. Rashedi. 2016. Particle size measurement in waste water influent and effluent using particle size analyzer and quantitative image analysis technique. Advanced Materials Research. 1133: 571-575.

Klenin, N. I., I. F. Popov and V. A. Sakun. 1986. Agricultural Machines. A. A. Balkema Publishers, Rotterdam.

Koster, K. H. 1985. Centrifugal discharge of bucket elevators. Bulk Solids Handling 5(2): 449 - 464.

Krivchenko, G. I. 1986. Hydraulic Machines, Turbines and Pumps. Mir Publishers.

Kulkarni, A. J., T. M. Kulkarni, O. J. Mahadik and P. V. Mahindrakar. (2018). Design of continuous loading vertical chain conveyor. International Journal of Advance Research, Ideas and Innovations in Technology. 4(6): 465-469.

Leitzel, R. E., and W. M. Morrisey. 1971. Air-float conveyors. Bulk Materials Handling. Vol. 1. (ed) M. C. Hawk. Univ. Pittsburgh, School of Mechanical Eng. 1971 pp 307 -325.

Levin, J. H. 1958. Unit handling of fruits and vegetables. Agricultural Engineering 39(9): 566 - 568.

Li, Jun-Xia, Xiao-Xu Pang and Yu-Jin Li. 2016. Research of dynamic characteristic of belt conveyor. Advances in Engineering Research (AER), volume 105: 304-312.

Mainwaring, N. J. and A. R. Reed. 1986. Mechanisms for gas-solids flows at low velocity in pneumatic conveying pipelines. Proceedings of the 11th Powder and Bulk Solids Conference. Chicago, May 1986.

Marcus, R. D., L. S. Leung, G. E. Klinzing and F. Rizk. 1990. Pneumatic Conveying of Solids. Chapman and Hall, London.

Marcus, R. D. and F. Rizk. 1985. The reliability of long-distance pneumatic transport. Conference on Reliable Flow of Particulate Solids. CMI, Bergen, August 1985.

Martinetti, A., L. A. M. van Dongen and R. Romano. 2017. Beyond Accidents: A Back-Analysis on Conveyor Belt Injuryfor a Better Design for Maintenance Operations. American Journal of Applied Sciences. 14(I):1-12.

Mazurkiewicz, D. 2014. Computer-aided maintenance and reliability management systems for conveyor belts. Maintenance and Reliability. 16 (3) (2014) 377–382.

Mazurkiewicz, D. 2015. Maintenance of belt conveyors using an expert system based on fuzzy logic. Archives of Civil and Mechanical Engineering. 15: 412-418.

Memane, V. S. and N. S. Biradar. 2015. Design and analysis of belt conveyor system of sugar industry for weight reduction", International Journal of Emerging Technologies and Innovative Research. 2(5): 1473-1477.

McGuire, P. M. 2009. Conveyors Application, Selection, and Integration. CRC Press Taylor & Francis Group, Boca Raton, FL 33487-2742. P. 1-2.

Mg Than Zaw Oo, Ma Myat Win Khaing and Ma Yi Yi Khin. 2019. Analysis of Belt Bucket Elevator. International Journal of Trend in Scientific Research and Development (ijtsrd), 3(5): 760-763.

Miller, W. F. 1958. Bucket elevators and auger conveyors for handling free-flowing materials. Agricultural Engineering 39(9): 552 - 555.

Mohsenin, N. N. (ed). 1965. Definitions and Measurements related to Mechanical Harvesting of selected fruits and vegetables. Pennsylvania Agricultural Experimental Station. Progress Report 257.

Mohsenin, N. N. 1970. Physical Properties of Plant and Animal Materials. Vol. 1. Structure, Physical characteristics and Mechanical Properties. Gordon and Breach Science Publishers. New York.

Mohsenin, N. N. 1972. Mechanical properties of fruits and vegetables - review of a decade of research applications and future needs. Transactions of the ASAE 15(3):1064 -1070.

Mohsenin, N. N. 1984. Physical Properties of Plant and Animal Materials. Vol. 1. (2nd ed.) Structure, Physical characteristics and Mechanical Properties. Gordon and Breach Science Publishers. New York.

Nafi, A., J.R.R. Mayer and A. Woźniak. 2011. Novel CMM-based implementation of the multi-step method for the separation of machine and probe errors. Precision Engineering, 35(2), 318-328.

Ng, K. L., L. A. Ang and S. C. Chug. 1982. A computer model for vibrating conveyors. Proceedings of the Institution of Mechanical Engineers. 200 (B2): 123 - 130.

Nicolai, R., J. Ollerich and J. Kelley. 2004. Screw auger power and throughput analysis. An ASAE/CSAE AIM Presentation Paper Number: 046134. Ottawa, Ontario, Canada, 1 - 4 August 2004.

Nicolai, R., A. Dittbenner and S. Pasikanti. 2006. Large portable auger throughput analysis. ASABE Annual International Meeting. Portland, Oregon. Paper Number: 066043.

Noyes, R. T. and W. E. Pfieffer. 1985. Design procedure for pneumatic conveyors in agriculture. ASAE Paper No. 85-3507, St. Joseph, MI.

Nwandikom, G. I. 1990. Yam resistance to mechanical damage. Agricultural Mechanization in Asia, Africa and Latin America (AMA) 21(2): 33 - 6

Nwandikom, G. I. and S. N. Asoegwu. 1990. Physical Characteristics of plantain fruits (*Musa AAB cv. Agbagba*). Proceedings of the NSAE 14:233 - 245.

Obiefuna, J. C. and T. O. C. Ndubizu. 1979. Estimating leaf area of plantain. Scientia Horticultura. 11: 31 - 36.

Oehman, H. H. 1981. Theory of vibrating conveyors. Bulk Solids Handling 1(2): 245 - 254.

Olanrewaju, T. O., I. M. Jeremiah, and P. E. Onyeanula. 2017. Design and fabrication of a screw conveyor.Agricultural Engineering International: CIGR Journal, 19(3): 156–162.

Palit, P. and A. C. Bhattacharyya. 1984. Measurement of leaf area per plant of shite jute (*Corchorus capsularis* L.) and tossa jute (*C. olitorius* L.) using the average specific leaf weight value. AGRIS 61(1): 59 – 62. https://agris.fao.org/agris

Perry, R. H. and D. Green. 1984. Perry's Chemical Engineer's Handbook. 6th Ed. McGraw. Hill, New York 7.11 to 7.13.

Peshatwar, S. V., D. Shubham, D. Prasad, D. Mahesh and G. Pratik. 2020. Review Design Analysis and Optimization of Screw Conveyor for Asphalt Application: A Review. Journal of Engineering Research and Application.10(01) (Series -II): 01-04

Pettingill, D. H. and W. F. Millier. 1968. The effect of certain design changes on the efficiency of a forage blower. Transactions of the ASAE 11 (3): 403 - 406, 408

Pinches, H. E. 1958. Materials Handling: farm production integrator. Agricultural Engineering 39(9): 517.

Puckett, H. B. 1958. A new automatic unloader for flat-bottom bins. USDA, ARS 42 - 17, May 1958.

Puckett, H. B. 1960. Machines to help with chores, In: Power to Produce (The Yearbook of Agriculture 1960), published by USDA pp 242 - 250.

Recommended Practice for Troughed Belt Conveyors. 1986. Mechanical Handling Engineers Association (MHEA), London, 1986.

Rehkugler, G. E. and L. L. Boyd. 1962. Dimensional analysis of auger conveyor operation. Transactions of the ASAE 12 (1): 98 - 102.

Roberts, A.W. and J.W. Hayes. 1979. Economic analysis in the optimal design of conveyors. TUNRA Ltd. Univ. of Newcastle, Australia.

Roth, L. O. and H. L. Field. 1991. An Introduction to Agricultural Engineering: A Problem Solving Approach 2nd Ed. Van Nostraud Reinhold, New York.

Rothwell, T. M., C. Vigneault and P. H. Southwell. 1991. Comparison of a pneumatic conveyor and a bucket elevator on an energy and economic basis. Canadian Agricultural Engineering. 33(2): 392 - 397.

Saker, A. A and H. Z. Hassan. 2013. Study of the different factors that influence jet pump performance. Open Journal of Fluid Dynamics, 2013, 3, 44-49.

Samuel, G.L. and M. S. Shunmugam. (2003). Evaluation of circularity and sphericity from coordinate measurement data. Journal of Materials Processing Technology, 139(1-3), 90-95.

Schertz, Cletus Earl, "Motion of granular material on an oscillating conveyor " (1962). Retrospective Theses and Dissertations. 2091.https://lib.dr.iastate.edu/rtd/2091.

Schroeder, S., S. Braun, U. Mueller, R. Sonntag, S. Jaeger and J.P. Kretzer. 2019. Particle analysis of shape factors according to American Society for Testing and Materials. Journal of Biomedical Materials Research Part B, 1-9.doi:10.1002/jbm.b.34382

Seferovich, G. H. 1958. Handling materials on the farms. Agricultural Engineering 39(9): 518 - 523, 537.

Segler, G. 1951. Pneumatic Grain Conveying. Braunschweig 1951

Singley, M. E. 1957. Self-feeder silo controls silage flow. Agricultural Engineering 30(2): 84 - 86, 93.65.

Singley, M. E. 1958. Handling non-free flow materials. Agricultural Engineering: 39(9): 540 - 542, 565.66.

Sitkei, Gyorgy 1986. Mechanics of Agricultural Materials. (Developments in Agricultural Engineering 8). Elsevier Science Publishers, Amsterdam, The Netherlands p.302.

Spivakovsky, A. and V. Dyachkov. 1985. Conveying Machines Vols I & II.Mir Publishers Moscow

Srivastava, A. K., C. E. Goering and R. P. Rohrbach. 1996. Engineering Principles of Agricultural Machines (Revised Printing). ASAE Textbook No. 6, St Joseph, MI. USA.

Strobel, J., J. Sumpf and H. Bankwitz. 2017. Dynamic Studies on a Slide Chain Conveyor System. International Symposium Plastic-Slide-Chains and Tribology in Conveyor Systems: Proceedings, Vol. 3, Chemnitz, p. 180-186

Sui, W. and D. Zhang. 2012. Four Methods for Roundness Evaluation. 2012 International Conference on Applied Physics and Industrial Engineering. Physics Procedia 24 (2012) 2159 – 2164.

Thayer, K. 2017. 10 Safety Precautions You Must Be Aware of When Performing Conveyor System Maintenance. GlobalSpec Engineering News. https://insights.globalspec.com/article/7518/. Assessed 10/12/2018.

Theint, Khema and Thae Mon Aung Thin Zar Lin. 2018. Design consideration of bucket elevator conveyor. IJIRST –International Journal for Innovative Research in Science & Technology 5(2): 1-4

Thomson, F. M. 1973. Applications of Screw conveyors. In: Bulk Materials Handling, Vol. II (ed). M. C. Hawk, School of Mechanical Engineering, Univ. of Pittsburgh p. 84 - 98.

Thompson, T. L., R. J. Frey, N. T. Cowper and E. J. Wasp. 1973. Slurry pumps: a survey. Proceedings of Hydrotransport 2, Bermuda Human Resources Association (BHRA) Conference, Conventry, U.K. September 1972, Paper, H 1.

Tomašková, M. and J. Sinay. 2014. The Use of Screw Conveyors in Practical Applications. Applied Mechanics and Materials. 683: 225–231.

Totten, D. S. and W. F. Millier. 1966. Energy and particle path analysis: Forage blower and vertical pipe. Transactions of the ASAE 9 (5) 629 - 636, 640.

TRAMCO. 2014. Centrifugal Discharge Bucket Elevator "The Workhorse" Assembly, Operation & Maintenance Manual. 44pp.

Udupa, G., M. Singaperumal, R. Sirohi and M.P. Kothiyal. 1998. Assessment of surface geometry using confocal scanning optical microscope. Mechatronics 8(3):187-215.

Vanitha, R. 2019. Conveyors Monitoring, Control and Protection Using Programmable Logic Controller. International Journal of Recent Technology and Engineering (IJRTE). 7(5S2): 250-254.

Wang, S., W. Guo, W. Wen, R. Chen, T. Li and F. Fang. 2010. Research on Belt Conveyor Monitoring and Control System. In: Zhu, R., Zhang, Y., Liu, B., Liu, C. (eds) Information Computing and Applications. ICICA 2010. Communications in Computer and Information Science, vol 105. Springer, Berlin, Heidelberg. https://doi.org/10.1007/978-3-642-16336-4_44

Wen, C. Y. 1971. Dilute and dense-phase pneumatic transports. In: Bulk Materials Handling Vol. 1(ed), M. C. Hawk, School of Mechanical Engineering, Univ. of Pittsburgh p. 258 - 287.

Woodcock, C. R. and J. S. Mason. 1987. Bulk Solids Handling (An Introduction to the Practice and Technology). Chapman and Hall, New York.

Zandi, I. 1982. Freight pipelines. Journal of Pipelines 2: 77 - 93.

Zareiforoush, H., M. Komarizadeh and M. Alizadeh. 2010. A review on screw conveyors performance evaluation during handling process. Journal of Scientific Review. 2(1): 55-63.

Zareiforoush, H., M. H. Komarizadeh, M. R. Alizadeh and M. Masoomi. 2010. Screw conveyors power and throughput analysis during horizontal handling of paddy grains. Journal of Agricultural Science 2(2): 147-157.

Zheng, M. Q. and Y. B. Hou. 2014. Application Research of Belt Conveyor Remote Maintenance Based on Cloud Computing. In Advanced Materials Research (Vols. 962–965, pp. 1132–1135).

INDEX

Printed in the United States
by Baker & Taylor Publisher Services